Salesforce CRM Administration Handbook

A comprehensive guide to administering, configuring, and customizing Salesforce CRM

Krzysztof Nowacki

Mateusz Twarożek

Salesforce CRM Administration Handbook

Group Product Manager: Aaron Tanna
Publishing Product Manager: Uzma Sheerin
Book Project Manager: Manisha Singh
Senior Editor: Nisha Cleetus
Technical Editor: Vidhisha Patidar
Copy Editor: Safis Editing
Proofreader: Nisha Cleetus
Indexer: Tejal Daruwale Soni
Production Designer: Shankar Kalbhor
Business Development Executive: Saloni Garg
DevRel Marketing Coordinators: Deepak Kumar and Mayank Singh

First published: April 2024

Production reference: 1190424

Published by
Packt Publishing Ltd.
Grosvenor House
11 St Paul's Square
Birmingham
B3 1RB, UK

ISBN 978-1-83508-569-1

www.packtpub.com

For my parents and all who've supported me along the way! Your support means the world to me.

– Krzysztof Nowacki

I would like to dedicate this book to my loved ones – my daughter Zosia, my wife Alicja, my parents, my sister and her family, my parents-in-law, and everyone who has supported me.

– Mateusz Twarożek

Contributors

About the authors

Krzysztof Nowacki is a Salesforce Certified Administrator and Consultant with 15+ years of experience in CRM and HR software implementations and consultancy for key international accounts in various industries (including FMCG, pharmacy, retail, manufacturing, finance, and SSC/BPO). He is also a speaker at conferences and a lecturer in postgraduate studies.

Mateusz Twarożek is a certified Salesforce administrator and consultant boasting 14 years of IT industry experience. Currently, he serves as the Head of Quickstarts for Salesforce Admins in the CEE region and acts as a consultant. His expertise spans system design in diverse sectors such as fintech, sales, manufacturing, and education, with a notable specialization in the NGO sector. As a conference speaker, he takes pleasure in disseminating his knowledge and insights within the community. Beyond his professional endeavors, he is a devoted spouse and parent who cherishes family time and has a personal interest in solving logical puzzles.

About the reviewer

Matt Thomas is a Salesforce Certified Professional with several certifications including Administrator, Data Cloud Consultant, and Business Analyst. As a consultant in different industries, he works with Salesforce CPQ, Sales Cloud, Service Cloud, Data Cloud, and Marketing Cloud. He holds an MBA from Carroll University and is a **Certified Treasury Professional (CTP)**. He enjoys hiking with his family, running ultramarathons, and attending U2 concerts with his wife.

Table of Contents

13

Continuing Education and Career Development   273

Preface

Greetings! Welcome to the world of Salesforce administration. The Salesforce CRM Administration Handbook is a comprehensive analysis of the functionalities of Salesforce that you need to know to become a Salesforce Certified Administrator. This book also includes practical tasks that you will be able to perform with us.

As you may already know, Salesforce is a platform that empowers businesses with its robust automation and management capabilities. In the realm of Salesforce, administrators play a pivotal role in overseeing the platform's functionalities, ensuring smooth operations, and driving business success. This book serves as your comprehensive guide to mastering Salesforce administration, specifically tailored to prepare you for the Admin exam.

Many resources cover features related to Salesforce, but in this book, we also focus on exam preparation and future career development. Additionally, we detail two of Salesforce's most popular products, Sales Cloud and Service Cloud, so this book serves not only as a resource for Salesforce Certified Administrator exam preparation but also as a first step toward obtaining additional Salesforce certificates, such as Sales Cloud Consultant and Service Cloud Consultant.

Drawing from years of experience in the Salesforce ecosystem and insights gathered from interviews with industry-leading Salesforce administrators and consultants with whom we have worked over the years, this book provides practical advice, best practices, and real-world scenarios to equip you with the skills needed to excel in Salesforce administration.

As per the latest analyst reports, Salesforce is a leader in the CRM space, with continued exponential growth projected in the coming years. With this growth, the demand for skilled Salesforce administrators is set to skyrocket. This book prepares you not just for the present but also for the future, ensuring you're equipped to handle the evolving challenges and complexities of Salesforce administration.

Whether you're embarking on a new career path or seeking to enhance your existing skills, this book is your roadmap to success in Salesforce administration. Let's dive in and unlock the full potential of Salesforce together.

Who this book is for

This book is for those embarking on their journey with Salesforce for the first time, aspiring to become Salesforce Certified Administrators in the future, and all Salesforce users who want to extend their current knowledge related to Salesforce.

The primary focus of this content is directed towards three key personas, namely the following:

People without hands-on experience with Salesforce – If you've ever thought about working in the Salesforce ecosystem but don't know where to start, then this book is for sure for you. It's sufficient to have a basic understanding of how the software works. The rest you will learn from us.

Salesforce beginners wanted to pass the Salesforce Certified Administrator exam – People with a few months' experience with Salesforce who want to prepare themselves for the Salesforce exam.

Salesforce junior consultants – These people will be able to complete their knowledge with this book, as we cover elements that can easily be overlooked in our day-to-day work.

What this book covers

Chapter 1, Getting Started with Salesforce, offers a thorough introduction to Salesforce, the premier CRM platform. Discover its powerful features for enhancing sales, service, and marketing processes. Learn about the Salesforce Admin role, essential skills, and the prestigious Salesforce Admin Certification Exam. We provide step-by-step instructions for setting up your account and efficient navigation techniques. Additionally, a comprehensive glossary helps you master Salesforce terminology.

Chapter 2, Salesforce Architecture, uncovers Salesforce's platform architecture in detail. This chapter covers topics such as multi-tenancy, understanding Salesforce Orgs' role in managing users and customization, and Salesforce instances and pods. You will gain insights for building a robust Salesforce environment.

Chapter 3, Getting to Know Data Management, offers a deep dive into Salesforce's data realm. It covers the data model, including objects and fields and their organizational roles. You'll explore data import/export techniques, mapping, and best practices for maintaining integrity. By the chapter's end, you'll master managing objects and fields, as well as importing/exporting data within Salesforce.

Chapter 4, Lightning Experience, covers the move from Salesforce Classic to Lightning Experience and how to personalize Lightning pages. Explore the benefits, plan for user adoption, and learn to create custom Lightning components. This chapter describes how to use the Lightning App Builder for page customization to gain skills to transition smoothly and enhance the user experience in your Salesforce Org.

Chapter 5, Objects in Salesforce, delves into the world of Salesforce objects to discover standard and custom objects, junction objects, and external objects. The chapter covers how to tailor Salesforce to your business requirements.

Chapter 6, User Management and Security, guides you through the intricate world of user permissions. The chapter discover the roles of user profiles, roles, and permission sets in defining access, explores sharing settings, organization-wide defaults, and field-level security to ensure data protection. You will also learn about login policies and two-factor authentication for enhanced security.

Chapter 7, Automation Tools, provides a comprehensive guide to mastering three critical aspects: approval processes, Flow Builder, and Apex triggers. This chapter brings practical insights and step-by-step instructions to streamline Salesforce customization. Additionally, the chapter will explore the AI features available in Salesforce.

Chapter 8, Reports and Dashboards, covers creating reports, report types, and building dashboards. The chapter gives insights and guidance on generating powerful reports and customizing them. The chapter also covers building interactive dashboards for data-driven decisions.

Chapter 9, AppExchange and Custom Applications, explores AppExchange, Salesforce's marketplace for pre-built applications. The chapter covers navigating AppExchange, selecting and installing apps effectively, and managing and maintaining apps.

Chapter 10, Service Cloud, provides a comprehensive guide to Salesforce's Service Cloud and delves into the intricacies of case management. The chapter covers key features and capabilities, as well as effective case management techniques, including case creation, assignment, escalation rules, and resolution best practices.

Chapter 11, Sales Cloud, provides a comprehensive guide to Salesforce's Sales Cloud, focusing on lead and opportunity management. This chapter explores key Sales Cloud features, including campaign management, lead capture, qualification, and opportunity management. The chapter also covers Account and Contact management and pipeline forecasting.

Chapter 12, Salesforce Administrator Exam Preparation, provides a comprehensive guide to preparing for Salesforce certification exams. The chapter offers insights into the exam format, scoring criteria, study strategies, and tips for exam day. This chapter helps you understand the exam structure, develop a study plan, and practice with sample questions. This chapter also covers the techniques for managing exam-day stress and maximizing performance.

Chapter 13, Continuing Education and Career Development, explores continuous education and career development in the Salesforce ecosystem. The chapter covers strategies for staying updated with Salesforce releases and leveraging resources. The chapter explores the benefits of joining the Salesforce community and offers insights into advanced certifications and career pathways.

To get the most out of this book

When reading a book about Salesforce, it's essential to have a basic understanding of the Salesforce platform and its fundamental concepts, such as data management, automation, and customization. Familiarity with IT application administration, support, and monitoring will also be beneficial for comprehending more advanced topics covered in the book.

Software/hardware covered in the book	Operating system requirements
Salesforce	Windows, macOS, or Linux
Apex	Windows, macOS, or Linux
Salesforce Lightning Web Components	Windows, macOS, or Linux

The Salesforce test environment, called Salesforce Developer Edition, should be created. This can be done at `https://developer.salesforce.com/signup`. Additionally, the reader should have access to the Salesforce Trailhead platform (accessible at `https://trailhead.salesforce.com/`), as access to both will be needed to perform the tasks described in the book.

> **Disclaimer on images**
>
> Some images in this title are presented for contextual purposes, and the readability of the graphic is not crucial to the discussion. Please refer to our free graphic bundle to download the images. You can download the images from `https://packt.link/gbp/9781835085691`.

Conventions used

There are a number of text conventions used throughout this book.

`Code in text`: Indicates code words in text, database table names, folder names, filenames, file extensions, pathnames, dummy URLs, user input, and Twitter handles. Here is an example: "The use of the `setPassword()` API for resetting passwords."

A block of code is set as follows:

```
trigger myLeadTrigger on Lead (before insert) {
    // Your code here
}
```

Bold: Indicates a new term, an important word, or words that you see onscreen. For instance, words in menus or dialog boxes appear in **bold**. Here is an example: "Place some action buttons on the record page such as **Edit**, **Clone**, **Delete**, and so on."

> **Tips or important notes**
> Appear like this.

Get in touch

Feedback from our readers is always welcome.

General feedback: If you have questions about any aspect of this book, email us at `customercare@ packtpub.com` and mention the book title in the subject of your message.

Errata: Although we have taken every care to ensure the accuracy of our content, mistakes do happen. If you have found a mistake in this book, we would be grateful if you would report this to us. Please visit `www.packtpub.com/support/errata` and fill in the form.

Piracy: If you come across any illegal copies of our works in any form on the internet, we would be grateful if you would provide us with the location address or website name. Please contact us at `copyright@packt.com` with a link to the material.

If you are interested in becoming an author: If there is a topic that you have expertise in and you are interested in either writing or contributing to a book, please visit `authors.packtpub.com`.

Share Your Thoughts

Once you've read *Salesforce CRM Administration Handbook*, we'd love to hear your thoughts! Scan the QR code below to go straight to the Amazon review page for this book and share your feedback.

`https://packt.link/r/1835085695`

Your review is important to us and the tech community and will help us make sure we're delivering excellent quality content

Download a free PDF copy of this book

Thanks for purchasing this book!

Do you like to read on the go but are unable to carry your print books everywhere?

Is your eBook purchase not compatible with the device of your choice?

Don't worry, now with every Packt book you get a DRM-free PDF version of that book at no cost.

Read anywhere, any place, on any device. Search, copy, and paste code from your favorite technical books directly into your application.

The perks don't stop there, you can get exclusive access to discounts, newsletters, and great free content in your inbox daily

Follow these simple steps to get the benefits:

1. Scan the QR code or visit the link below

https://packt.link/free-ebook/978-1-83508-569-1

2. Submit your proof of purchase
3. That's it! We'll send your free PDF and other benefits to your email directly

Getting Started with Salesforce

In this chapter, we provide a comprehensive introduction to Salesforce, the leading CRM platform. We'll explore the features and capabilities that make Salesforce a powerful tool for streamlining sales, service, and marketing processes. Delve into the Salesforce Admin role, understanding the primary responsibilities and essential skills required for success. Learn about the prestigious Salesforce Admin certification exam, including its format, the covered topics, and the benefits it brings to your career. Discover how to set up your account with step-by-step instructions, ensuring a smooth entry into the platform. Dive into the interface as we guide you through efficient navigation techniques, allowing you to explore Salesforce's features and functionalities effortlessly. Furthermore, master the Salesforce terminology with a comprehensive glossary that demystifies key concepts and empowers you to communicate effectively within the platform. By the end of this chapter, you'll have the tools and knowledge needed to confidently embark on your Salesforce journey and unlock its full potential.

In this chapter, you will learn about the following topics:

- Starting with Salesforce
- The Salesforce Admin role – primary responsibilities and needed skills
- The Salesforce certification exam
- Setting up your Salesforce account
- A tour of Salesforce – getting around the interface

Starting with Salesforce

We have always tried to record all kinds of information, sometimes to immortalize it, other times to show it, but most importantly, to not forget this information.

Let's start with the main question – what is CRM? Its software, named after the acronym derived from customer relationship management, helps companies manage customer interactions. Every business needs a database where it can record the data of its contractors and their history, transactions, and price lists. This used to be recorded in notebooks and notepads, later in Excel spreadsheets (some companies still use them as their main customer database), and now, one of the leading CRMs in the world is Salesforce.

Is Salesforce like a crystal ball that will boost sales, and phone calls and emails will make themselves? No! (Although emails can be sent automatically.) But thanks to CRM, businesses will be able to do the following significantly:

- **Serve the customer better**: Automation will do all the necessary tasks for us, remind us of a forgotten customer, and free employees from repetitive actions that always take up precious time and divert attention from the key elements of the business – such as relationships.

- **Understand the customer better**: The system enables understanding and anticipating the needs of customers, generating sales trends, and most importantly, taking care of them and the business's relationship with them.

- **Increase customer engagement**: With tools that allow tracking of customers and their interactions with the business, personalizing content directed to them, and knowing their preferences and tastes, businesses are able to get to know their customers. Consequently, businesses can direct and adjust their business strategies to the style of their customers.

- **Achieve better sales results**: As we have already pointed out, Salesforce is not a crystal ball predicting the future, but if we know our customers, then those customers know us and our product. And that is already another step toward achieving excellent sales results.

"Well, give me a CD with this system and let's install it," some customers might say. But Salesforce is not software that we install on a server or local disk; it is a cloud solution. Thanks to this, everyone can have access to their data.

Salespeople can check their customers during business trips, the marketing team can check the results of their campaign during an event, and the service technicians can make necessary notes from anywhere in the world. The system is a multi-layered ecosystem that is scalable and meets the needs of the customer. What does this mean? This means that customers who want to use sales support elements do not have to purchase the entire solution along with marketing and service desk support.

The available system functions are included in individual licenses. Purchasing a Salesforce license is a subscription model. The customer purchases a specific type of license that offers a certain range of functionality for a period of 12 months.

The prices of the license at the time of writing for the Sales Cloud range from $25 to $300.

Salesforce servers are located in many parts of the world such as the United States, Canada, Europe, Latin America, Asia, and Australia.

Is customer data safe? Of course; Salesforce is certified for compliance with many security standards, including ISO 27001, SOC 2, and FedRAMP.

The Salesforce solution is directed to customers of different sizes. From greenfield-type customers who have no previous experience with the system, to huge corporations that expand their systems to enormous sizes, whose administration and maintenance are handled by large, specialist IT teams. Currently, this cloud solution is used by over 150,000 customers in 190 countries. Among the customers using Salesforce, we will find companies such as Spotify, Amazon, Canon, Toyota, Walmart, Uber, and many others.

Salesforce is a multi-module tool. This means that it has many different modules dedicated to specific needs, which will now be described briefly, and in the following chapters, you will learn much more about them:

- **Sales Cloud** is a solution tailored to the needs of commerce. With it, you can manage the sales cycle from acquiring a lead to winning an opportunity and selling products/services and much more. This module, thanks to its built-in tools, enables the use of forecasting and numerous reporting tools.

- **Service Cloud** is a tool dedicated to every service/help desk. Everyone has had to call a hotline to get help with a particular matter of life at some point. It's highly likely that your data has been recorded in Service Cloud. Thanks to its structure, it stores your data and your query in cases.

- **Marketing Cloud** and **Marketing Cloud Account Engagement** is a marketing pair that will manage marketing aimed at B2B or B2C. Automated emails and marketing paths are just a fraction of what you can achieve in the software of the marketing duo.

- **Commerce Cloud** is nothing more than a booster for your online store. It helps manage online stores, both for B2B and B2C. It personalizes product recommendations and facilitates promotion management, and can do much, much more.

- **Experience Cloud**, formerly known as Community Cloud (for many, this name will forever remain in their hearts), allows you to create pages, portals, forums, help pages, and others for customers, partners, or users. The drag-and-drop editor facilitates page management for administrators.

Are these all the clouds? No. There are many solutions that facilitate work in various business sectors. Thanks to this, Salesforce easily adapts to even the most complex tasks and needs.

> **Tip**
> Get a basic understanding of every Salesforce Cloud, its core features, and how it is helping the business to fulfill its goals, as questions related to this topic may appear on the Salesforce Certified Associate exam.

You already know that there is a Salesforce, how powerful it is, and how much it can bring to the life of any business. However, it is worth remembering that every system needs its own custodian, someone to look after how the system works and make changes to it. You will find out about this person in a moment – the Salesforce Admin.

Exploring the Salesforce Admin role

Every system requires a custodian, a person who can implement changes in structure, fix bugs, or create a new user. Although many of these changes can be automated, a human will always be necessary for system management or even its setup. That's what a Salesforce superhero – a System Administrator – is for. The Admin role is extremely important, but it may differ across systems. Many companies may define the responsibilities of this role differently. Sometimes it involves system maintenance, sometimes user management, and sometimes deploying previously created system solutions. The main task of Administrators is to manage and adapt the system to the needs of the customer. Thanks to these changes, the system should also meet the business goals of the organization. System Admins often combine technical knowledge with the use of soft skills when in contact with the client.

So, whether it's changes in the system, creating automation, or custom reports, the Salesforce Admin role is key to business. It happens that other roles, such as SF Developer or SF Solution Architect, perform administrative duties, but these are scenarios in which employees perform tasks dedicated to another position.

In conclusion, the Salesforce Admin is a key role for a business. Thanks to this role, the optimized system works better and more efficiently, and users can enjoy more efficient processes and a user-friendly structure. In the following points, we will look more closely at the main duties of the Administrator and what skills the "ideal" Sys Admin should have.

Main responsibilities of a Salesforce Admin

Once, Peter Parker (AKA Spiderman) heard from his uncle: "*With great power there must also come great responsibility.*" It is the same with a System Admin. Remember, you are the company superhero and you need to help them. Here are your responsibilities.

User management

As mentioned earlier, administration equals user management.

But it's not just about creating new users, but also about elements closely related to the user: License, Profile, Role, Permission Set. System Administrator is a role in which one of the main tasks is creating new profiles and assigning them to specific users. Knowing what necessary access users should have, the specialist can create a profile dedicated to them and assign it during the creation of new access permissions. If the profile does not meet all requirements, it can be supplemented with a Permission Set, in which access can be granted in a trimmed manner and given to individual users. When setting up new access to the system, it is important to check the license to be granted and the profile. If the company uses a hierarchy, the user should also have a role assigned to them.

But what would user management be without creating, updating, or deactivating users? With appropriate access, the System Admin can make these kinds of changes. Fun fact! You can't remove a user from the system; the only option is to deactivate the user by unchecking the "Active" checkbox.

Roles/permissions management

Salesforce is a software in which roles and permissions play a huge role. As mentioned earlier, the main element granting access is profiles, but what if a given department, consisting of two groups, needs a default set of access, but also two separate access permissions data dedicated to them? Then the System Administrator creates a Permission Set, which is a set of access permissions to data elements, objects, or features in the system.

Assigning roles, on the other hand, plays an important role in the structure of the company, thanks to which the administrator will build a hierarchy of employees in Salesforce.

How often have you found data in your system such as Jane/Joe Doe or QWERTY? Often, users want to enter something quickly using so-called "dummy details." That is, they enter incorrect data to quickly create a record in the system. And it is the System Administrator who stands guard over the integrity of the data (imagine them standing on a high building with a flowing cape and a large "A" on their chest). The Sys Admin must ensure that the data entered into the system is correct, unique, and above all, true. Salesforce has the appropriate tools that help achieve data cleanliness, such as the following:

- Validation rules that check the correctness of data entered by users
- Duplication rules that regulate the possibilities of creating (or not creating) records with the same data
- Approval processes, which are requests for data acceptance on the record (e.g., the amount of discount on the opportunity)

To achieve high cleanliness and integrity of data, the Administrator will use the full range of tools offered by Salesforce.

But what if it turns out that users are exceptionally creative and find a way to create duplicated data and introduce Mr. QWERTY 321321321? Then, a red code lights up above the Admin's head and a data audit needs to be carried out. Such audits may include checking for the following, among other things:

- Duplicates (e.g., 2 x Mrs. Jane Doe with the same email)
- Data inconsistencies (e.g., Acme Corp and Acme Corporation as two different accounts)
- Data errors (in the text field – e.g., email: `edward.scissorhands@acme` without `.com`)

The Admin should be able to detect these types of errors, correct them, and set up the appropriate safeguards that will act preventively. Regular data cleaning will keep the database clean and accurate. As we all know, a good admin is the best friend of any department where mass import/export is used. This is also one of the essential skills that an admin must master. For such actions, they can use Data Loader, Data Import Wizard, or Salesforce Inspector. Are any of these particularly recommended? Each Administrator or person dealing with mass data management has their own style and favorite tool for these tasks.

> **Tip**
> If you want to efficiently input data into the system, prepare upload templates for your users and teach them to fill in data in the correct structure. Remember, you are the Salesforce specialist, not your users. A correct data structure in the upload template will save you a lot of time.

Customization

This mysterious-sounding word is one of the most important in the daily tasks of a Salesforce Admin. It should be remembered that Salesforce in the out-of-the-box version is a tool that is not tailored to the client's vision at all. Opportunities have their default stages, on Account – apart from the website and telephone, you will also find a fax field, and the contacts do not contain many data fields.

This is where the administrator, who, like Michelangelo, sculpts Salesforce for business needs, steps in. Customizing the system is an extremely important process, which can last for either of the following:

- **Periodically** – in the form of projects
- **Constantly** – as part of system management

System configuration can contain many elements, from changing branding, adding new fields, and changing settings, to creating completely new home page layouts. These tasks are often repetitive, such as creating a field or removing it, but among the tasks, there are also those that require more time and practice, such as automation. When Salesforce announced the retirement of Process Builder and Workflow Rules, fear befell people who effectively avoided Flow, the only native low-code tool currently available in the Salesforce ecosystem. Thanks to its user-friendly interface, Flow has become popular not only with Administrators but also with other people working on improving the company's org. It is important to remember what I mentioned at the beginning: each company follows its own policy and scope of work for Administrators. So, in one business, it will only be structural changes, and in another, it will be automation and deployment between environments.

Training and troubleshooting

Once the system is configured and populated with data, what's left for the Administrator – to sit back and enjoy the lack of tasks? Of course not.

Every company hires new employees, many of whom might not have worked in the industry before, and an even larger number may not be familiar with Salesforce. This is another task assigned to the Administrator: it's up to them to guide users through the ABCs of Salesforce.

The training can be broken down into the following segments:

- **Salesforce basics**: Logging in, creating records, inputting data into fields, creating custom reports, and so on

- **Company processes in Salesforce**: An introduction to sales/service processes, essential data in records, set approvals/validations, automation, and so on

> **Tip**
>
> If you want users to remember something from your training, ask questions, show examples on the existing org, and remember that you are speaking to users unfamiliar with SF. Therefore, adjust your language and try to explain everything from the basics.

Often, users, even unknowingly, will encounter a scenario that wasn't tested during the solution design. They then look to specialists for help. With various solutions such as ticketing tools, emails, or service cases, the Sys Admin can efficiently address user issues. Typically, resolving user issues involves checking their profiles, permission sets, and general object settings. Communication with users is crucial in such situations, allowing the Admin to replicate the problem and immediately solve it.

A highly useful feature when addressing user issues is the **login as user** option available under **Setup -> User -> Login** next to the user:

Figure 1.1: Login as user

With this, the administrator can view the system from the user's perspective, enabling a quicker resolution to the problem at hand.

> **Tip**
>
> If you don't want to be logged out every time after using **login as user**, change the session settings in **Setup -> Session Settings -> Deselect Force re-login after Login-As-User**. This way, the system won't force a re-login after you switch back.

Skills

Working in the Salesforce environment is not solely based on knowledge of the system.

Many beginner Administrators have never dealt with this system before. Maybe they've heard of it but never had any experience with it; many of them will have studied cultural studies, sociology, computer science, or agriculture. Is this an obstacle to starting work as an Admin? No. What matters are willingness and knowledge of SF (or a great desire to get to know it), and it's good to have analytical, problem-solving, communication, and project management skills (e.g., agile skills).

However, it's worth noting that many companies offer internships for Administrators and various academies that prepare future Administrators. Quite recently, Salesforce launched a program for women who want to return to work after a break related to motherhood, illness, or other personal reasons, and their goal is the SF environment. The program is called "Bring Women Back to Work," and as the name suggests, it is exclusively for women. After a 12-month training cycle, SF helps these women enter the job market by sending their resumes to program partners.

Do you need to have all these skills to be a good Admin? I think they are not must-haves, but rather nice to have, because with them and the right knowledge, you can achieve the title of Certified SF Admin. There's more on this in the next section.

Salesforce certification exams – what can you expect?

To be certified or not to be certified? That is the question! I often receive this query from Salesforce newcomers seeking advice on establishing a career in Salesforce. The straightforward response is, "*Yes, certainly, you should obtain Salesforce certifications.*" The more considered answer is, "*It depends.*" You'll frequently come across a common description on many LinkedIn profiles of Salesforce experts – "2x Salesforce Certified," "5x Salesforce Certified," "21x Salesforce Certified" – showcasing their number of Salesforce certifications. Should you follow the same path? Yes, you can. Is this the only path? No, it is not. Allow me to elucidate this using an analogy my jiu-jitsu trainer shared with me during my first tournament, when I found myself facing a rather intimidating individual covered in many tattoos, who happened to be my next opponent. My trainer's words were, "*Tattoos do not fight.*" This statement holds true universally and can be applied to the Salesforce realm as well. Essentially, having a multitude of certifications doesn't always equate to genuine proficiency in Salesforce.

Throughout my IT career, I've encountered numerous Salesforce professionals possessing "just" one or two certifications who were exceptionally skilled in Salesforce. To put it simply, practical experience always outweighs certifications – always. However, this doesn't negate the value or importance of holding Salesforce certificates. Allow me to outline some pros and cons concerning obtaining Salesforce certifications.

Pros

There are several benefits associated with obtaining a Salesforce certificate and even attempting to acquire one:

- When preparing for a Salesforce exam, you will need to study to know more about Salesforce
- You will learn about topics and features that you may not have used previously so you will increase your knowledge
- You will get a nice "tattoo" that others will recognize and it will be visible as proof of your knowledge
- It may help you jumpstart your Salesforce career
- If you are a Salesforce beginner, having certs such as Admin and/or App Builder will strongly increase the chance of you getting your first Salesforce job
- More is not always better but sometimes, having the next Salesforce cert will give you hints about which Salesforce career direction to take or not take

Cons

Let's now see the cons related to exam preparation:

- You will study things you may not even touch/use outside the exam.
- The knowledge you will gather may be forgotten quickly if you do not use it in practice.
- More certs do not mean your career will rocket quicker as experience > certs, and in most recruitment processes, you will be asked about your practical knowledge.
- Specialization is still strongly looked for in the IT world. The market is not searching for individuals who are A-OK in every single Salesforce feature/cloud. The recruiters are searching for experts in particular features or clouds.
- Doing certs for the sake of having certs is wrong. Be sure that the knowledge you will get may be used practically and is in line with your career plan.

Of course, the aforementioned "cons" are not real cons, as having a certificate is almost always better than not having it, but before planning your next Salesforce cert, you need to answer an important question: what do you want to do in your life? And then, just…take the relevant exam.

Analyzing Salesforce exams

We will cover the Salesforce exam topic more deeply in *Chapter 12*. Here, I will only highlight the most important topics related to the Salesforce exams in general:

- *There are more than 20 Salesforce-related certificate types on the market now*

  Yes, some people have them all. No, more does not mean better.

- *The exams are multiselect tests*

  The exception is the Salesforce Associate Certification, which is currently not multiselect.

- *The passing score may differ but is generally around 65%*

  You should know that some topics require a higher score than others as their wages are higher.

- *You may do the test online at home or offline in one of the exam centers*

  As the exam centers may not be available in your area, you may always take the online exam. The online exams are proctored, so this means that you will be assisted and monitored by a proctor through a webcam.

- *There are official test questions provided by Salesforce*

  Currently, test questions are available only for Admin and Email Specialists. You can buy these official test questions to prepare yourself for those exams.

- *If your native language is not English, you may ask Salesforce to increase the time you will need to deal with the exam*

  You will get additional time if you are taking an exam in English and your native language is not English. To get additional time, you need to contact Salesforce support before the exam.

- *To get some certificates, you may need to achieve other certs as some are prerequisites for other exams*

  For example, you need to be a Certified Administrator to be able to get an Advance Administrator certificate.

- *There is a clear path that you can follow to become a Salesforce Certified Technical Architect, the top of the top roles in the Salesforce ecosystem*

  Some Salesforce certs are bound together and passing them all will ultimately give you the title of Application Architect and/or System Architect, which opens the door to becoming a Salesforce CTA – a Certified Technical Architect. The paths are known. Just follow the steps.

As I mentioned already, the Salesforce Certified Administrator exam will be covered in detail in another part of this book, so I would like to mention only that the Admin exam was, up until some time ago, the first exam that opened the doors to the Salesforce cert world. But as the exam scope is quite big and it's better to be at least a Salesforce user on a daily basis to pass this exam, Salesforce some time ago introduced an easier exam called the Salesforce Certified Associate. Salesforce just recognized that it's hard for Salesforce newbies to prepare for the Admin exam and start their careers without having at least a few months experience on the platform using Salesforce in the current work. The Salesforce Certified Associate is dedicated to Salesforce beginners with around three months' experience on the platform, while the Admin cert is recommended to people with around six months' Salesforce experience. The first three exams we recommend doing to jumpstart your Salesforce career are as follows:

- Salesforce Certified Associate

- Salesforce Certified Administrator

- Salesforce Platform App Builder

You may ask, do I really need to do the Salesforce Certified Associate exam before trying to pass the Admin exam? The answer is simple: yes but no. Just joking. Yes, you may skip the Associate exam and go straight to the Admin exam. This is recommended especially if any of the following apply:

- You are starting an SF-related job soon – let's be brutally honest, the Salesforce Associate exam is a good start but when you have already got a Salesforce job, in 99% of cases that means you know some Salesforce and you would need more skills than are needed from Associate. Rather, you would need at least Admin skills. This is because the Associate is rather one repair wrench and when you start a Salesforce-related job (especially in consulting), you need the whole toolbox already.

- You are already a Salesforce user/superuser/Admin – if you are already in one of these roles, the Associate exam could be tempting as you will pass it without needing to learn it, but it will add almost nothing of value to your career. Of course, you will be "more" certified and will be able to make the "huge" change on LinkedIn, which will be changing "1x Certified" to "2x Certified" on your profile summary, but mostly, that's it.

- You learn fast or you've worked before with other IT tools such as CRM, ERP systems, and so on.

- You have limited funds – Salesforce Associate is not free on a daily basis. Sometimes, Salesforce gives some random discounts that can limit the price to zero or cut the cost, but usually, you need to pay for the exam. If your funds are limited, you may consider skipping the Associate exam as the Admin certificate is much more recognizable and valued than the Associate cert.

Tip

You may already know that there is also another popular Salesforce certificate that is dedicated to programmers. It's called Platform Developer 1. Some may say that you don't need to do certs such as Admin or App Builder if you just want to be a programmer. They would be wrong. Salesforce is a platform where one of the most important rules is to try to not use Visualforce… Just kidding. The real rule is "try to configure before starting to code." As Salesforce gives you a lot of customization possibilities, code-based automation options the starting point should always be declarative customization, click no code aka low code. For this reason, I recommend starting your Salesforce certificate journey with Admin, then App Builder as a prerequisite before jumping to Dev 1, even if you don't want to be an Admin and do not know what App Builder really is.

Now that you have some basic context about Salesforce exams and being Salesforce certified, let's see how to make the very first step toward your future Salesforce career and learn how to create your Salesforce test account.

Setting up your Salesforce account

Is it even possible to learn Salesforce without working in Salesforce at your job? This question is asked many times by Salesforce Admin wannabes. The answer is: yes, yes, yes!

I like to say that the Salesforce job market has quite a low entry threshold when comparing it to other popular IT platforms such as SAP, Oracle, Microsoft CRM, and so on. This is because of a few important reasons:

- You may learn Salesforce on your own, at your own pace, using Salesforce's free learning platform, Trailhead.

- Salesforce exams are not too expensive – this is of course a relative statement, but when comparing Salesforce exam prices to other technologies' exam prices, the Salesforce ones are not as pricey.

- In addition, you can get a fully useful and free test environment that you can use to get to know the platform and customize, build, and destroy safely. In this section, we will show you how to do this.

Free Salesforce environments

There are two types of free Salesforce environments: Developer Edition orgs and Trailhead Playgrounds. Let's see what the differences are and how to create each one.

Free Developer Edition environment

"What? A developer environment? Should I use it, given that I'm not a developer and have no intention of becoming one?" The answer is straightforward: yes, you can use it even if you are not a programmer. Think of it as a name that Salesforce has assigned to its demo account. To see a more detailed explanation, let's take a look at how Salesforce describes its Salesforce Developer orgs on their pages: "A Salesforce Developer environment grants you access to a fully featured copy of Force.com, absolutely free, and yours to use indefinitely. Utilize the Developer Organization to stay current with the latest features, construct your own applications, and test functionalities prior to their purchase." This implies that you can access the Developer Edition for free, forever and ever. It encompasses both Sales Cloud and Service Cloud features, enabling you to explore and experiment with both sales-related and customer-support-related functionalities within a single space.

Can you generate multiple Developer Edition environments? Yes, you can. Will you create more than one? Yes, you certainly will! Based on practical experience, I can confirm that you'll utilize dev orgs for customizable experimentation, constructing numerous automations, preparing job interview demonstrations, and a wide array of other tasks. And you will do it a lot. For these reasons, it's often preferable to initiate with a fresh new environment rather than attempting to "repair" the alterations you made in your existing orgs.

Free Trailhead-related Playgrounds

A long, long time ago in the Salesforce galaxy far, far away, there existed an era when the only testing environments that Salesforce provided to users were the free Developer orgs. However, with the advent of Salesforce Trailhead, a free learning platform, the demand arose for quicker and more convenient access to testing orgs. This prompted the creation of Trailhead Playgrounds to address this requirement. Essentially, Trailhead Playgrounds are akin to Developer Environments, but they can be generated with greater ease, speed, and relevance to specific Trailhead tasks.

> **Tip**
> I use Salesforce Dev environments for client demos, job interview demos, testing new Salesforce releases, and doing some experiments, while I use Trailhead Playgrounds only when doing Trailhead-related tasks.

I hope you now understand the difference between the Salesforce Developer environment and Trailhead Playgrounds. Let's now see how to create them.

Let's start with the Developer environment. These are the steps to create a Salesforce Developer environment:

1. Go to the Salesforce Developer website:

    Visit `https://developer.salesforce.com/signup`.

2. Fill in the required information:

    Fill in the required data such as your name, email address, password, and other details. Agree on the terms and conditions and click on the **Sign me up** or **Create Org** button at the end:

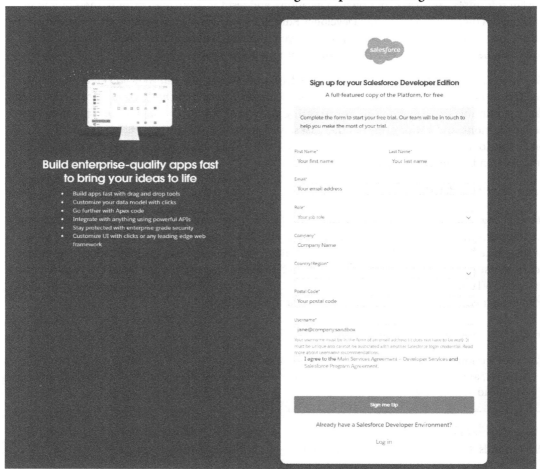

Figure 1.2: Fill in the form

3. Verify your email address:

    After registering, Salesforce will send you an email to verify your email address. Click on the verification link provided in the email:

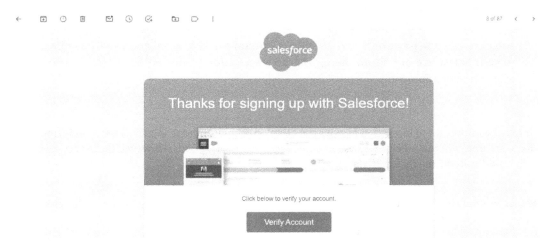

Figure 1.3: Verify account

4. Complete the registration process:

    Once your email is verified, follow the instructions to complete the registration process. Set the password that will be used to access the Developer Account:

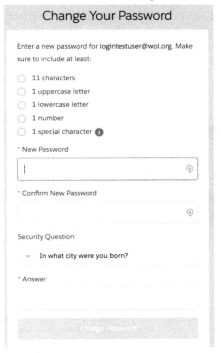

Figure 1.4: Set a password

5.   Create a new Developer Edition:

After creating the password, your Developer Edition will be created and you'll be redirected to your Salesforce org. From there, you can start building and testing applications in the Salesforce environment.

Tip

It may be important to know that besides the Developer environment, which as was said is free for a lifetime, you may also get access to other Salesforce tools/clouds. However, those accesses are time-limited (mostly for 30 days).

Now that we know how to create a Developer environment, let's see how to create a Trailhead Playground:

1.   Sign up or log in to **Trailhead**:

If you don't have a Trailhead account, go to the Salesforce Trailhead website (`https://trailhead.salesforce.com`) and sign up for a free account. If you already have one, log in using your credentials.

2.   Access Playgrounds:

Once you're logged in, navigate to the **Hands-On Orgs** section of the Trailhead website (`https://trailhead.salesforce.com/users/profiles/orgs`). This is where you can manage your Trailhead Playgrounds:

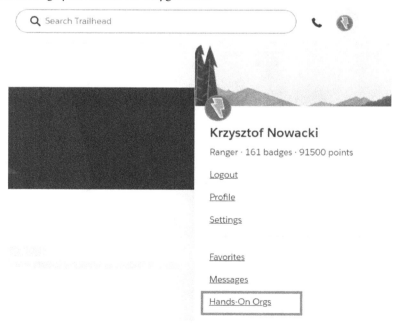

Figure 1.5: Hands-On Orgs

3. Create a new Playground:

    Look for an option that allows you to create a new Trailhead Playground. Click on it to begin the process:

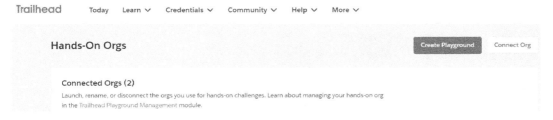

Figure 1.6: Create a Playground

4. Choose a Playground name:

    Give your new Playground a unique name that helps you identify it easily. The name should be relevant to what you plan to use the Playground for:

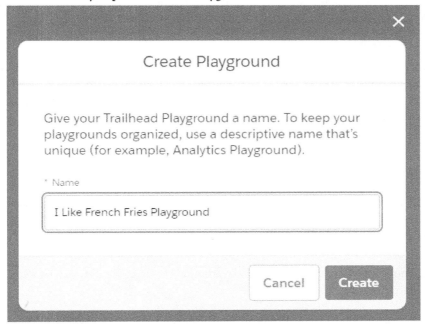

Figure 1.7: Choose a Playground name

5. Build the Playground:

    Once you've filled in the required details, click on the **Create Playground** button. The system will then create a new Salesforce Trailhead Playground for you:

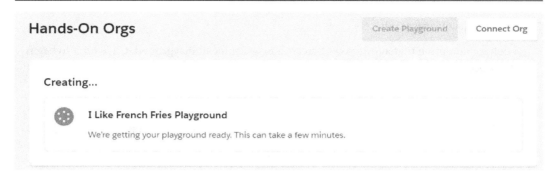

Figure 1.8: Build a Playground

6. Launch the Playground to access it:

    After the creation process is complete, you'll receive information about your new Trailhead Playground including its unique URL, credentials, and any other relevant details.

    With your new Trailhead Playground set up, you can now access and explore Salesforce features, try out different configurations, build applications, and practice what you've learned in the Trailhead modules.

7. Create Playgrounds ad-hoc:

    Besides creating "in-advance" Playgrounds that may be then used during hands-on Trailhead challenges, you may also create a separate/new Trailhead Playground for each Salesforce Trailhead hands-on challenge that you will be currently performing. While taking the test challenges, you will see the action button that will serve you with ad-hoc Playground creation.

> **Tip**
> Although the way to create Salesforce Dev orgs and Trailhead Playgrounds has not changed for some time, you need to remember that the process might change or evolve over time, and Salesforce may introduce new steps or change the user interface. Always refer to the official Salesforce Developer website or documentation for the most up-to-date information on creating a free Developer Edition environment.

Now that you know how to create a Salesforce Playground, let's see in the next section of this chapter what the Salesforce interface looks like.

A tour of Salesforce – getting around the interface

In this section, we will closely examine the Salesforce user interface, delving into its most essential components. This will provide you with a better understanding of what you, as a user, will encounter. If you are reading this book, it's likely that you have intentions to dedicate a significant amount of time to navigating the Salesforce UI. As a result, grasping the features linked to the Salesforce UI becomes of utmost importance.

Comprehending these pivotal elements of the Salesforce UI will empower you to use the platform more efficiently, thereby enhancing your overall productivity. Keep in mind that the Salesforce UI is both intuitive and adaptable, allowing for customization to cater to individual preferences. This, in turn, maximizes the potential of Salesforce, facilitating success in your business endeavors. By investing time in familiarizing yourself with these features, you can fully leverage Salesforce and harness its capabilities to drive your business forward.

In this section, we will cover the following Salesforce tabs, also known as objects:

- Home
- Leads
- Accounts
- Contacts
- Opportunities
- Reports
- Dashboards

We'll also cover the following Salesforce features:

- Search Bar
- List Views
- Setup access
- Object setup access
- Record setup access

Home – where your Salesforce journey begins

The main home page after logging in is a crucial starting point. Here, you will find a concise overview of important information, such as tasks, calendar events, notifications, and other key data pertinent to your work. By the way, the Home tab has a really good name. I will tell you why. Because like with a real home, you will abandon it after a while, spending more time in other places such as bars, gyms, and restaurants – in other words, other Salesforce tabs such as **Lead**, **Account**, **Contact**, or **Opportunity**.

> **Tip**
> If you are a Salesforce Admin, speak with your business to understand what they would really like to see on the Salesforce home tab. If you do not do this, they will leave the house (home) and will never miss it. ;)

The four riders of significance – Leads, Accounts, Contacts, and Opportunities

Moving on, the top navigation bar contains various tabs and options, facilitating seamless navigation within the Salesforce platform. This enables swift access to sections that interest you, such as **Contacts data**, **Customers Data**, and **Prospects data**.

- **The Lead**: The person initially interested in your product or service.
- **The Accounts**: Companies such as your customers, partners, future customers, and competitors.
- **The Contacts**: People related to Accounts, most likely employees of Accounts. People working for your customers, partners, future customers, or competitors. People who you like. People who should like you too. ;)
- **The Opportunities**: Sales processes processed with your customers, partners, and future customers. Deals that you love. Money that they will bring. ;)

How are Accounts, Contacts, and Opportunities bonded together? Let's see a simple example to be able to understand the relations between Accounts, Contacts, and Opportunities in Salesforce. Let's map example customer data from the company "Super Sweets" to Salesforce features; let's "take us to the candy shop":

Account – "Super Sweets" candy shops

Contact – Adam Smith, who works for Super Sweets as an IT Director

Opportunity – Selling IT security stuff to Super Sweets and contacting Adam Smith to make it happen

So, you will keep the Super Sweets company information such as name, address, and customer status on the Account object, while people working at Super Sweets will be covered by the Contact objects. Deals that you are making with Super Sweets will be tracked on the Opportunity objects. Quite straightforward, isn't it?

Leads

Sales users will most likely embark on their Salesforce journey right here. The term "Leads" refers to individuals we intend to reach out to, as they could potentially express interest in testing our solution, purchasing our product, or booking our service. A Lead embodies elements of an Account, Contact, and Opportunity; it represents a person (with attributes such as name, email, and phone number akin to Contact fields) associated with a company (with a **Company Name** field resembling an Account) that we wish to evaluate for potential interest in our offerings (indicated by the Lead's status, similar to Opportunity indicators).

Why does the Lead exhibit such similarity to Account, Contact, and Opportunity? This alignment stems from the fact that Leads serve as precursors to these entities, as all their respective records can be generated during a procedure called "Lead conversion." So, what precisely is Lead conversion? Allow me to elucidate using an example.

Imagine your company specializes in high-end merchandise. Someone completes an online form inquiring about your products and provides all the necessary contact details. This individual now qualifies as a Lead. It's that straightforward. Your sales team can now reach out and ascertain the extent of their interest in the products. If, following a conversation, the salesperson confirms their continued interest in our high-end merchandise, the Lead can be elevated beyond its initial status – it becomes what you might call an "uber Lead." The salesperson can initiate the conversion process, transforming the Lead into an Account, Contact, and Opportunity – all three records emanating from that single Lead! This process is aptly termed Lead conversion.

Throughout Lead conversion, fields from the Lead, such as **Company Name**, facilitate the creation of Account records, while attributes such as first name, last name, and email are employed to generate Contact records. When the Lead is transformed into an Account and Contact, the Opportunity can also be concurrently established. Opting to create an Opportunity during Lead conversion signifies the continuation of the sales journey, presenting an opening to potentially sell a product or service to the Lead.

Wondering how to trigger Lead conversion? The process is both prominent and straightforward. An action button labeled **Convert** exists on the Lead page. Simply click this button and work your magic!

Reports and dashboards

After a while, you will notice that your Salesforce database is growing, and after some time, you will realize that there is a substantial amount of data. Eventually, your business will also recognize this and inquire about reports that can present crucial information concerning Leads, Opportunities, Orders, and more. Be prepared! Form a friendly relationship with reports and dashboards in the Salesforce Analytics tab.

So, what exactly are reports and dashboards in Salesforce? It's actually quite straightforward. A report is used to create a summary of existing Salesforce data. It bears some resemblance to an Excel spreadsheet, displaying data in columns and rows. On the other hand, dashboards are designed to visualize this data. Think of them as Excel graphs but integrated within Salesforce.

Reports don't necessarily require dashboards to function; they can stand alone and remain independent. They possess strength and autonomy. However, dashboards cannot exist without reports. Dashboards are more like team players; they are often associated with numerous reports that serve as their data sources. They possess strength as well, but their power is rooted in the reports they are built upon.

Search Bar

Now, what can this feature be used for? Can you guess? ;) Well, of course, it's meant to help you locate the closest bar in your neighborhood. Just kidding! That's a task expertly handled by Google Maps or your best friend! In reality, Salesforce employs the Search Bar exclusively for locating the records you're seeking. Just enter the name of a record, such as an Account Name or Contact Surname, and Salesforce will promptly provide you with relevant results. It's as straightforward as that.

What is searchable in Salesforce?

- Standard object records such as Leads, Contacts, Accounts, Opportunities, Orders, and so on
- Custom object's records if they are marked as searchable in their settings
- Reports and dashboards

How to use search?

- Just write a word, click **Search**, and get records. Magic! ;)
- Use natural language while searching (with some limitations, of course), for example:

  - `my contacts lead source is partner`
  - `opportunities in Texas last month`
  - `tasks high priority last week`

- Use wildcards and operators such as the following:

  - asterisk (`*`) to find items that match zero or more characters at the middle or end of your query
  - question mark (`?`) to find items that match only one character at the middle or end of your query

- Use search operators such as the following:

  - `AND` – search for items that match all of the search terms
  - `AND NOT` – search for items that don't contain the search term
  - `OR` – search for items containing at least one of the search terms
  - Parentheses – group search terms together (grouped search terms are evaluated before other terms in your string)
  - Quotation marks – search for items that match all the search terms in the order entered

> **Tip**
>
> Salesforce updates its search capabilities from time to time for this reason. Try to keep yourself up-to-date and check the current Salesforce release and help pages related to this subject. For example, the AI-related and natural language search is what Salesforce has developed strongly in recent years. If you are using the Salesforce Enterprise edition, you can catch a glimpse of these features when searching and activating Einstein Search in your setup.

List Views

What is the Salesforce List View? Let me explain this with a simple example. Have you ever noticed that when entering Salesforce **Accounts**, **Contacts**, or any other tab, Salesforce serves you a list of recently seen records? I hope you noticed this because the recently viewed records are an example of the Salesforce List View. List views are lists of records that could be created on every tab.

The main features related to List Views are as follows:

- There are standard List Views such as **Recently Viewed** but besides those, you can also create custom list views.

- You can decide which Salesforce fields should be used as columns on the list while displaying this to the users.

- You can set filters to show only records that you want, for example, only records owned by you or records with specific field values, such as Accounts with the billing country as Poland or Spain.

- You can decide who sees the created list views. Different (public) groups, users, or users with specific roles can have their own List Views that are visible only to them, so as to not overcrowd the list views of other users. Finally, you may also create list views that are private, so visible only to you.

> Tip
> Please consider restricting the permission to create a public list view (yes, such a permission exists in Salesforce and can be revoked from certain users) to Salesforce Admins and Super Users. Granting this capability to every individual user will lead to chaos in the public list views, causing an influx of unnecessary views and user frustration, ultimately resulting in users abandoning this tool. Instead, let Salesforce Admins or business-oriented Super Users, who are closely attuned to the needs and requirements of standard users, manage the list views.

Related lists

Related lists are very similar to List Views, but they always exist in the context of records only. What are Related Lists, then? To describe them, let's use simple examples of standard related lists:

- List of Contacts related to a specific Account
- List of sales Opportunities related to a specific Account

Both aforementioned related lists can be seen on the Account records UI and show people and deals related to the specific Accounts. While viewing Account X, you will see the list of Contacts and deals related to Account X, while, when viewing Account Y, you will see the Contacts and deals related to Account Y. As simple as that!

The main tasks of Related Lists are as follows:

- Show records related as a child records of the record you're currently on

- Be able to show filtered records; when using dynamic related lists, you can show only a portion of related records limited by some filter conditions

- Create new child records with the press of a button, just by clicking the **New** action on the specific related list

- Perform some mass actions on the records displayed on the lists, such as updating or deleting those records

Setup access and direct object setup access

The possibility of accessing the setup means that you are the master of the universe, or at least the master of your Salesforce realm. You are quite literally the "boss" of this digital "neighborhood." To enter the Salesforce setup, all you need to do is click on the gear wheel icon located in the top-right corner of the screen:

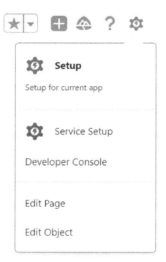

Figure 1.9: Setup access

From this perspective, you can enter the whole treasure chamber by clicking **Setup** (or **Service Setup** if you want to mess around with Salesforce Service Cloud settings), or **Edit Object** if you want to go directly to one of the "treasure chambers" such as **Accounts setup**, **Lead setup**, and so on. How will you know which "chamber" you are entering? It always uses the context of the tab that you are currently in. So, if you are in the **Account** tab and you click **Edit Object** there, you will land on the **Account Setup** page. The same goes for other Salesforce tabs.

What can you do there? Many interesting things, such as the following:

- Add new fields
- Create field validation
- Create record action buttons and list view buttons

Record UI setup access

You might have already noticed that when you are on a specific record page and you click the gear wheel icon, you will find not only the **Setup** and **Edit Object** options but also the **Edit Page** option. What is this option for, you ask? Well, it's a Salesforce mystery! Just kidding! While there are indeed many Salesforce mysteries out there, this isn't one of them. When you click the **Edit Page** option, you'll be redirected to the area where you can manage the UI or interfaces related to the object of the record you're currently viewing. This includes the UI of objects such as Accounts, Leads, Opportunities, and more:

Figure 1.10: Object access

What can you do in this setup? Many interesting things, such as the following:

- Decide which fields are visible on the record's UI
- Decide whether some fields should be visible, for example, only when other fields have some specific values
- Place some action buttons on the record page such as **Edit**, **Clone**, **Delete**, and so on
- Place some components on the record's UI such as paths, related lists, tabs, rich text, and many, many more
- Place custom-made (coded by Salesforce programmers) components

Summary

In summary, this initial chapter has given us the basic understanding needed to grasp why Salesforce is important and how it works.

We've explored the pivotal role of a Salesforce Admin, gaining insights into their key responsibilities and the essential skills they need to excel. Now you know what a Salesforce Admin does, so you will now consciously follow this challenging path and you won't be saying that no one warned you about it. So, no excuses from now, sorry!

Additionally, the section on the Salesforce certification exam has illuminated the path to becoming a certified professional. Now you know that more is sometimes less, and that you don't need to be a Salesforce General with tons of badges and certificates on your chest to become successful, desired, or who knows, maybe even loved by the market.

Practical guidance has been offered on setting up a Salesforce account, followed by a detailed tour of the platform's interface, enhancing readers' ability to navigate and utilize Salesforce effectively. You are now able to create a free and endless Salesforce environment, one or a hundred of them. OK, let's say even thousands of them. You will stop counting after the tenth.

We hope you enjoyed the first chapter of this book and are eager to explore more! Remember that Salesforce is all about development, and here, we are talking not only about app development but also, most of all, about self-development, where we believe perseverance is a key skill. So, *"Stay hungry, stay foolish,"* – Steve Jobs and remember, *"We've (all) suffered losses, but we've not lost the war."* - Optimus Prime. This is just the beginning, and the real deal will come in the next chapters! Buckle up! Here we go!

2
Salesforce Architecture

In today's era, technology plays an unimaginably crucial role. We encounter it daily, holding a mobile phone in our hands, staring at a laptop screen, or even validating a ticket in public transportation. Salesforce is also a technology that propels businesses into their digital renaissance and streamlines their processes. However, behind the sleek interface lies a complex system architecture, allowing users to operate within the Salesforce ecosystem. In this chapter, we will dive deep into the sea of technology on which the system is built, helping us understand its structure and discerning why it functions the way it does. We'll cover the following topics:

- Understanding multi-tenant architecture
- Discovering Salesforce orgs – what does it mean?
- Salesforce instances and PODs

Understanding multi-tenant architecture

In this section, you will learn about the types of system architectures. You will understand the difference between the following three types of structures:

- Multi-tenant
- Single tenant
- Hybrid systems

You will get to know the challenges faced by landlords when establishing boundaries between one client and another and making updates. Most importantly (remember this is a book about Salesforce), you will find out why the leading CRM uses a multi-tenant architecture.

Multi-tenant structures

In the world of technology, "multi" is quite a common term: multi-tasking when talking about phones, multi-tenant when discussing cloud solutions, and multi-universe in the context of superhero movies. Oh, wait – we will not be focusing on that last part in this book.

When we mention multi-tenant, we are referring to a model where a single software instance serves multiple clients. All these clients use the same code base, resources, and infrastructure but are isolated from each other and operate independently.

Multi-tenant can be likened to a hotel; let us call it the Dark Side of the Moon. Hotel guests arrive and use the same plumbing, service, building, and communal areas, but each has their own room, accessible only to them and the management.

This model is frequently adopted by hosting companies and cloud solution providers. It offers various advantages. Easy software updates and scaling are possible while simultaneously catering to many clients.

However, security is paramount for the client. Using the hotel analogy, what would happen if you received an access card but mistakenly went to the wrong floor? Guided by habit rather than checking the number, you would find yourself entering and settling into a bed already occupied by John Doe. (A story based on true events.) I assume you would be very surprised, and more importantly, you would never patronize that hotel again. The same principle applies to multi-tenant structures; companies uphold stringent security standards, creating barriers between one client's data and another's. The next diagram is a simple explanation of what such a structure looks like:

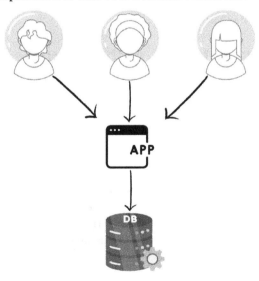

Figure 2.1: Multi-tenant architecture

Why am I delving into this? Because Salesforce, regarded as a leader in cloud technology (they do have a cloud in their logo; come on!), employs multi-tenancy. This enables them to provide services to millions of clients while maintaining high quality, efficiency, and security. But we might wonder what would happen if Salesforce changed its structure, and each org became a separate server. I think I can describe it by quoting an ABBA song: *Money, Money, Money...*

So, how does Salesforce implement such a solution? Salesforce utilizes advanced resource allocation and monitoring technology to ensure that every tenant can use their designated number of resources, such as computing power, memory, and storage. This technology uses flexible resource throttling and load balancing. This is why Salesforce can dynamically adapt to tenants' requirements while guaranteeing real-time computing power for all.

Salesforce has been perfecting this multi-tenant technology for many years, achieving levels of performance, scalability, and security previously thought unreachable.

How to attain such a structure? Naturally, it is not simple. Every tenant receives a unique ID.

Each ID is subject to stringent access controls. When a user logs in to the system, they are verified for the correct username and password, assigned to a specific org, and given access to the ID. This ensures that the user only views their data.

But how can a tenant customize their space? Let us go back to the "Dark Side of the Moon" hotel example. What if the staff did not allow us to bring in and unpack our luggage, limiting us to just those small shampoos? That would not be convenient. That is why almost every software operating in a multi-tenant structure provides the ability to customize its software to meet client needs. The first and most crucial element is isolating the configuration and personalization of each client, allowing them to tailor their environment to their vision of the ideal CRM. The second is deeper customization, such as adding their components. Salesforce has achieved a very high level in this field, as the system allows for the use of Apex, Visualforce, or Lightning Components. And the last is integration, connecting the environment with external applications.

Yet, what would modern technologies be without challenges? For some, it is defeating the main boss on a console; for others, it is preparing a meal; for multi-tenant technology, the challenges include resource management, conflicts between tenants, updates and modifications, and compatibility of custom solutions.

Let us discuss each of these, starting with resource management. In our example hotel, there are many guests, but what if one guest took all the hot water? The remaining guests would be displeased. Therefore, Salesforce equitably distributes its computing power. Every tenant gets exactly as much as the others.

Conflicts – well, there probably is not a person in the world peaceful enough to never have had a conflict. In hotels, various situations can arise that may affect other guests. For example, the unfortunate Mr. Doe caused a short circuit, resulting in a power outage on the entire floor. As a result, Ms. Doe could not connect her laptop to the scheduled meeting and missed the opportunity to sell an excellent consulting project – a bit of a butterfly effect, right?

Similarly, in Salesforce, certain actions can affect others. However, currently, Salesforce closely monitors everything on the orgs, ensuring such situations rarely occur.

Tip

If you want to find out if something is currently happening with your instance or if there is any maintenance, visit `https://status.salesforce.com/`. There, you will find all ongoing incidents or maintenance.

Updates are the next tasks to do so allow me to paint you a brief scene. Imagine a married couple, Jane and John Doe, spending their time at our hotel, Dark Side of the Moon. Just as they are settling down to sleep, there is a knock, and hotel staff enter wanting to change the door lock. Sounds inconvenient, doesn't it? The same goes for updates; they must be well timed.

That is why if you are familiar with Salesforce, you've likely heard about the new releases. They are announced well in advance, and before such an event, you receive a notification upon login that the org will be briefly unavailable. Thanks to these releases, we gain new functionalities and safeguards.

And now, the final piece of the puzzle is the compatibility of custom solutions. As you probably know, our hotel, Dark Side of the Moon, is quite accommodating to its guests. One day, a certain Vlad, specifically Vlad Dracula, arrives at the hotel from distant Romania. Given that Mr. Dracula is not fond of sunlight, he requests the installation of sun-blocking blinds. Since Vlad is a long-standing, indeed a very long-standing, guest, the hotel agrees and installs special blinds. It is the same with our multi-tenant structure. Tenants can customize their systems to their needs, but there is one rule – it must be compatible with the main version of the system.

But are there other solutions that users employ? Indeed, there are.

Single-tenant structures

Besides the multi-tenant structure, there are also models such as single-tenant and hybrid models. In the single-tenant model, a tenant utilizes only the resource dedicated to them. This ensures full infrastructure-level isolation, a significant advantage in terms of security and performance. However, it also entails additional costs. Individual resources must also be updated individually or by the tenants themselves.

Hybrid structures

Then, there is the hybrid structure, a blend of shared and tenant-dedicated resources. Such a structure provides a balance between isolation and costs. Updates can be tailored to the client's needs but also incorporate all the crucial updates. Next, you will find an illustrated diagram of the two architectures – hybrid and single-tenant:

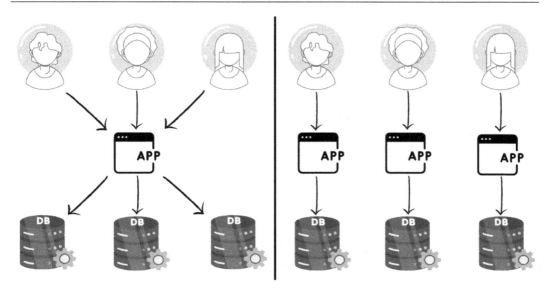

Figure 2.2: Hybrid versus single-tenant architecture

Salesforce has chosen the multi-tenant structure, built its solution's strength on it, and executed it well. Thanks to this, customers are pleased with the quality versus cost balance.

Of course, the choice of the appropriate architecture depends on the needs of a particular client. However, when deciding, it is crucial to understand the benefits and limitations of each.

Owing to its cost-effectiveness, ease of management, and high scalability, the multi-tenant architecture has become a key element in today's cloud technologies. Still, one must remember that every solution has its pros and cons: single-tenant, multi-tenant, or hybrid, each has its challenges.

When considering the implementation of a specific solution, one must think through each of these models and check its adaptability to their needs. What matters isn't the model in which the new solution will be implemented but that the world is moving forward, and innovative technologies are opening the door to exciting new journeys. Thanks to this chapter, you now understand the structures that exist in the world of cloud solutions. And most importantly, you know the style in which our CRM is designed. The next time a client asks you, you can confidently explain how users connect to Salesforce. And how they create orgs. Wait, wait – do you already know this term? No? Then see you in the next section, which will clarify this term. See you later, administrator!

Discovering Salesforce orgs – what does it mean?

In this section, you will learn about the concept of an org, find out where it comes from, and discover the types of orgs used by Salesforce. Did you know that there is something called the lifecycle of an org? No? Well, you are about to find out. In the end, I'll provide you with best practices that will save you from many, many questions from your users and other administrators.

As you probably already know, Salesforce has its own universe – its structure and even its language are tailored to it, and believe me, Salesforce sounds best in English. But, getting to the point, many times when you have spoken, are speaking, or will be speaking to people working within this ecosystem, you will hear phrases such as "*in my org*" or "*but my client's org is truly a work of art.*" You might be wondering what this whole "org" really means.

Well, "org" is short for "organization." But do not equate it with a regular company or organization as we know it in our daily lives. Salesforce uses "org" to denote a unique environment centered on a set of data, its users, and its configured functionalities.

In the Dark Side of the Moon Hotel, there are rooms. These are our orgs, and they can have different standards, different curtains, double or single beds, among other differences (one of them has Vlad in it, remember!). Similarly, orgs are housed on Salesforce servers, with each serving as a distinct system holding its tenant's data.

Each org is an independent ecosystem where administrators and users operate. They log in to it day after day, working without hindering users of another org on Salesforce. Within this enclosed space, the people responsible tailor the system to their needs, creating processes and objects, analyzing data, and connecting it to other external systems.

So, remember – an org is not just an organization. An org is a unique environment tailored to a company's needs, where records, configurations, or custom solutions are stored.

Remember how I described the several types of rooms in our hotel? Double or single beds, a jacuzzi on the terrace, or maybe just a shower? Each room varies in its amenities. The same applies to orgs.

We have several types of orgs that cater to various needs throughout the life cycle of an application or configuration. By getting to know these types, you will learn how to effectively manage and deploy solutions in Salesforce:

- **Production org**: This is the primary environment where administrators log in, but also users who deal with Opportunities, convert leads, or create reports. It contains real data and business processes. If you delete a field from a Contact, the change will be immediately visible.

- **Sandboxes or test orgs**: Perhaps you're familiar with *Star Wars: The Clone Wars*? It works similarly here – we clone our org. These are clones of the production environment primarily used for testing and further system development. Salesforce offers several types of sandboxes:

  - **Developer sandbox**: This environment is for minor changes. It only contains metadata from the production org.

  - **Developer Pro sandbox**: Essentially, a more advanced version of the previous type. What is the difference? Simply, it provides more storage space for data.

  - **Partial Data sandbox**: This is a sandbox for special tasks. It contains metadata and a subset of data from the production org. It is perfect for testing.

- **Full Copy sandbox**: This one is a powerhouse. It is a complete copy of the production org. It includes all data, metadata, configurations, and development. It is best suited for integration tests.

- **Developer org**: This environment is utilized by developers. They test their solutions here. This environment allows for the creation and testing of new solutions. They are not tied to the production org and are a separate environment containing so-called dummy data or fictitious data. Developer org was mentioned in the first chapter, in the *Setting up your Salesforce account* section.

- **Partner org**: This environment is accessible to Salesforce partners. Such a partner can request an org that will be available to them, including for testing. Salesforce partners can request multiple such environments with different clouds.

How to create a new sandbox? Well, you can do it all in Salesforce **Setup**.

Let us do this together; follow the next steps, and you will set up your first sandbox with ease:

1. Log in to your org with administrator access.

2. Go to **Setup**, and in **Quick Find**, type sandboxes and click on the item that appears in the search results, visible in the following screenshot:

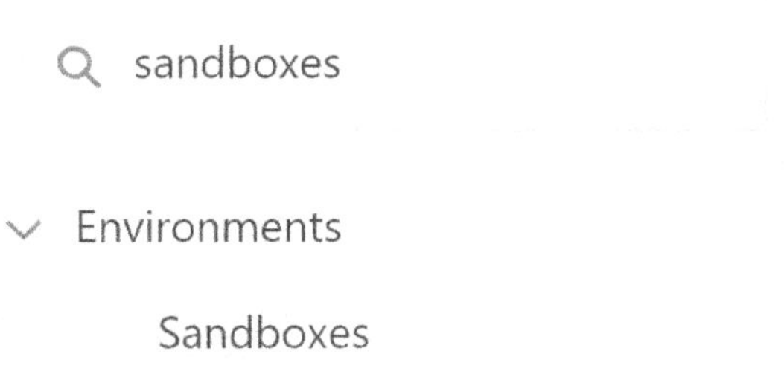

Q   sandboxes

∨   Environments

Sandboxes

Figure 2.3: Sandboxes in Setup

3. Click the **New Sandbox** button shown in the next screenshot:

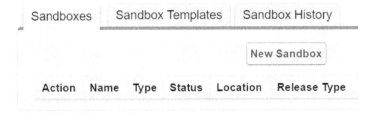

Figure 2.4: New Sandbox

4. Choose the type, enter the sandbox name, and select the data you want to have in your sandbox, as shown in the following screenshot:

Sandbox Information		▌ = Required Information
Name	▌ [ ]	
Description	[ ]	
Create From	[Production ∨]	

Figure 2.5: Sandbox types

5. Choose sandbox access – either **Public User group/All Active users** – and click **Create**.

As you can see, Salesforce has several types of orgs. By understanding the characteristics of each type, you will be better able to choose the org that fits your needs.

So, now you know what an org is and its types, but what is it physically made of?

- It has several important components, such as the following:

- **Data space**, which is of interest to every client. Typically, when starting any implementation, clients ask if all their records will fit there – the usual answer is yes (of course, provided the number of these records does not exceed 10 million). It's important to remember that in every org, data is stored in relational databases. These contain information on customers, transactions, and all custom data created by the company – this refers to objects. It's worth mentioning that these objects have relationships with each other to provide the user with comprehensive access to a client's entire history.

- **Configuration**, and here we come in – administrators! Salesforce enables customization of the org to meet a company's needs. You don't have to be knowledgeable in programming to fully rearrange the org according to your employer's requirements. Modifications to processes, fields, objects, accesses, and a million other options are readily available (actually just two clicks away: **Gear Icon -> Setup**).

- Once the org is customized, we can't forget about those who will use the system – our **users and their permissions**. Administrators usually handle adding new individuals and assigning them permissions. Salesforce, with its sophisticated permissions system, allows for granting and revoking access to the minutest elements of the system. Regarding permissions, the most crucial tools for an administrator are profiles and permission sets.

- And the last one, **feature access**, which is about smart license management and appropriate system configuration. Thanks to these two elements, users have access to the relevant features and apps in the system. It's worth noting that apps in Salesforce can be out of the box, meaning default, or custom-built by administrators. (An app is a set of objects visible on the top tab.)

Salesforce bases its architecture on the org concept, and to fully understand it, we need to grasp its significance for several types of users.

The significance of an org for users is that it is not only a place to work, but it is also the cycle of their tasks, needs, and duties. It is daily work. The Salesforce org frequently serves as the central hub for employees' work activities. They carry out the majority of operations and activities related to their clients there. A well-configured system is invaluable for day-to-day tasks, and customized processes can automate many duties for the users. The next significance is constant access to data. Thanks to visibility settings, users can see records that, for example, belong to them or their group. This eliminates the time they would have to spend filtering the right results from a massive database. Importantly, users can also access their records through the Salesforce mobile app, which reduces errors when copying numbers or unclear client summaries. Last but not least, customized access to applications. When configuring a new org, administrators set appropriate access levels – to records, fields, objects, or apps. As a result, a user will only have access to the apps that contain objects specifically dedicated to them.

However, it's not just users who operate in the digital Salesforce world. Let's explore the significance of orgs for administrators. The coolest part is configuration. The org is the central location where changes can be made. Thanks to the ability to create custom objects, it can be tailored to any requirements. Being a Texas ranger, this is security. This means controlling user access. Tools related to permissions, profiles, and permission sets allow for this. And the last one is managing key aspects of the system. In addition to the aforementioned two, the administrator also manages the system's performance, its accessibility, and its integration with external systems.

Salesforce is a central hub for work and management for both roles. Complementing each other, they form a cohesive mechanism that brings appropriate benefits to the company.

The lifecycle of a Salesforce org is a specific process that has certain stages. These stages are crucial and must be adhered to during implementation. Of course, also remember that if you have your developer org or if you work on a client's org that is pristine (such clients are called greenfield), you can make changes in the production environment.

> **Tip**
> Remember that after each implementation, before UAT, and then before the Go Live date, clear the org of records used for testing. All the John/Jane Doe, Fester the Tester, and Test12345 must be deleted.

The lifecycle of an org is somewhat similar to the human life cycle. Everything begins with birth. And it ends – well, I probably don't need to add how it ends. It is worth remembering that to work with Salesforce is to enter this cycle of org and follow its flow:

1. **Org creation**: We already covered this topic at the beginning of this chapter. We create a production org (such an org can be created by Salesforce for a client, or a client can request such an org via the website).

2. **Modification and development**: Changes of this kind are typically made in sandbox-type orgs. There, modifications and improvements are introduced as business needs evolve. These changes do not affect the production environment thanks to a separate sandbox.

3. **Testing**: At this stage, another actor comes into play, and that is QA. This individual conducts tests of all implemented solutions, based on previously created user stories (functionality descriptions).

4. **Change deployment**: Once the changes are approved by QA, it's time to move them to the production environment. With tools such as Change Sets or third-party solutions, these changes don't have to be moved manually (that is, reconfiguration in the production environment) but are transferred automatically and seamlessly.

5. **Org deletion**: It's time to say goodbye (tears might well up). It's not a common scenario since sandboxes are used for a longer time for solution development. However, sometimes a given sandbox is no longer needed and can be deleted. But it's crucial to be cautious in such actions because it's an irreversible process.

To the preceding points, we can add those that will surely raise the level of org management and its quality. After each implementation, documentation is crucial. It's important to include all changes made to it. This makes it easier to understand which changes were made and why. Another advantage of creating documentation is to pass on the knowledge to another person – imagine you get a new job, and a week before you leave, another administrator arrives; handing over documentation and a brief introduction will be much easier than lengthy meetings.

After introducing changes, besides proper communication, users need to be taught how to use the new functionalities. Training ensures users are aware of the solutions introduced in the last deployment. This training assures the user that it's a new feature, not a system error or bug.

But remember – there will always be someone who comes to you with questions, and the famous "turn it off and on again" might not work here. So, calmly introduce the new functionality for the 148th time to the user who didn't get it during the training.

Post-deployment feedback: Once everyone knows how to use the new solution, send them a survey and let them comment on the new functionality. Sometimes, the unique approach of users can outline a different concept of this functionality that we might not have thought of.

I sincerely hope that "org" will not remind you of Shrek but of Salesforce. As you can see, the concept of this architecture is essential to us, those associated with this cloud solution. With knowledge about the lifecycle of an org, you can plan your implementations in the correct sequence. Throughout your career, you will frequently encounter concepts of org, sandboxes, or perhaps multi-tenant structures, but thanks to what you've read, you now know what it's all about. But have you ever heard about PODs? No – we're not talking about a P.O.D. band here, but something closely related to Salesforce. The next section will explain what we define as a POD.

Salesforce instances and PODs

In this section of the chapter, you will discover the fascinating world of Salesforce PODs. But what is a POD? The acronym POD stands for Point of Deployment, which may sound a bit enigmatic. To better understand this, let's start with something you may have heard of – the word "instance." Salesforce PODs are a different name for Salesforce instances. Instances or PODs are the places where your Salesforce data is located. It is a server location where your Salesforce org is hosted and from where it operates. Does the instance equal the org? No! On the Salesforce instance, there can be multiple Salesforce orgs allocated. Of course, each Salesforce org is separated and secured, so you can't access the information of orgs that you don't have access to. However, orgs from different companies are located in the same instance. As you may already know, Salesforce has thousands of customers, so you may now be wondering if all of them are located in the same instance. The answer is no! Salesforce has many POD/instance locations. The separation between different instances is made geographically, so there are different instances across North America, Europe, and Asia.

In 2023, the Salesforce-managed data centers for the aforementioned services are located in the following metropolitan areas:

- **Chicago, Illinois, United States (USA)**
- **Dallas, Texas, United States (USA)**
- **Frankfurt, Germany (DE)**
- **Kobe, Japan (JPN)**
- **London, United Kingdom (UK)**
  - **London North**
  - **London West**
- **Paris, France (FRA)**
- **Phoenix, Arizona, United States (USA)**
- **Tokyo, Japan (JPN)**
- **Washington DC, United States (USA):**
  - **Washington DC North**
  - **Washington DC South**

In addition, Salesforce has instances served from Amazon Web Services (AWS) cloud infrastructure in the United States, Canada, India, Japan, Indonesia, Singapore, Brazil, Germany, Italy, Sweden, Great Britain, Korea, and Australia. These instances are in two or more separate Availability Zones within each respective country.

As you can see, Salesforce instances are located worldwide. The reason for this is strongly related to technical aspects but also legal ones. Some countries have just law regulations saying that the data related to their citizens needs to be stored in their region/country. Does it sound easy? Let's add some complexity because apart from country differences, Salesforce instances can be configured on first-party infrastructure such as the following:

- Infrastructure owned and operated by Salesforce
- Public cloud infrastructure: Hosted on Salesforce-managed AWS infrastructure (excluding Hyperforce)
- Hyperforce infrastructure: Hosted on Salesforce-managed AWS Hyperforce infrastructure

Typically, a customer has just one org to support all services dependent on it. In simpler terms, your org will be hosted on either Salesforce's first-party infrastructure, public cloud infrastructure, or Hyperforce infrastructure, but not on more than one of these simultaneously. Additionally, some Salesforce services may include functionality that interacts with the org but is hosted on separate infrastructure, such as Einstein features.

To summarise, Salesforce PODs or instances are simply the locations where your Salesforce org is situated. Physically, these are servers situated in data centers located in various regions around the world. So, even though we often use the phrase "cloud data" in relation to Salesforce, the data is not actually stored in the real clouds but in some familiar, solid cities such as Berlin, Chicago, or Kobe. Not as romantic, is it? But look – you might be fortunate, and your data could be located in Paris. Maybe it's not in the cloud, but there's no denying that Paris is equally romantic. And what about Phoenix, Arizona? Still not in the cloud, but at least it's in "The Valley of the Sun," right? Want to know how to check where your data is located or when you'll get new Salesforce features? We'll discover this in the next section.

Where is my Salesforce org located?

You may now be wondering, "Where is my Salesforce org located?" or "Is it working correctly?" and "Is there any planned maintenance?" There is a quick way to answer this question. As knowing where the data is really located is one of the most important security- and law-related topics, Salesforce provides us with a way to find this information.

To check what is your Salesforce org instance, just follow these simple steps:

1. **Get your instance**: To get this, navigate to **Setup -> Company Information**. Check the value of the field instance. It may look like this: EU44. Here's a screenshot example:

Organization Edition	Enterprise Edition
Instance	EU44
Modified By	<u>Krzysztof Nowacki</u>, 24.10.2023, 14:54

Figure 2.6: EU instance

2. **Go to the following Salesforce page**: `https://trust.salesforce.com/`. After entering the page, navigate to the **Status** tab, and in the search bar search for your instance name:

Figure 2.7: Salesforce Trust search instance

3. **Check the information related to your instance**: There is a lot of interesting information that you may get from the Salesforce status page, such as the following:

Current status by service: Here, you can check the current availability status related to Salesforce services such as core features, Search, Service Cloud, CPQ, Einstein, and many more.

- **History:** Here, you can check the availability history.

  For example, here you can see that on September 8, there were some issues with the instance (you can identify marked issues by looking for an exclamation mark placed on the timeline):

Figure 2.8: Incident

To get more info about the incident, just click on it and read the description. You will also see the start time, end time, and the duration:

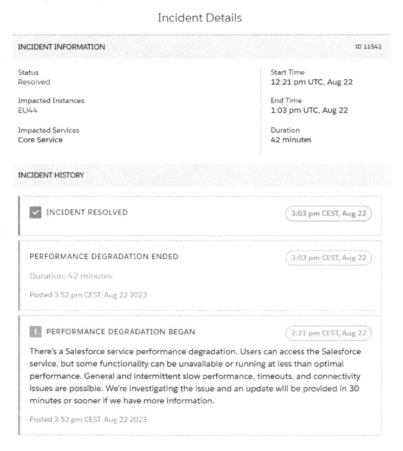

Figure 2.9: Incident details

- **Maintenance:** A list of historical and future maintenance related to this instance:

	INSTANCE
	EU44

CURRENT STATUS   HISTORY   MAINTENANCE

4/16/24 - 12/31/25 • 12 Items   Expand All | Collapse All   Q Quick Find

ID	DATE	STATUS	START TIME	SUBJECT	INSTANCES	SERVICES	TYPE
PAST 33 DAYS (1)							▲
682328	Apr 07	Confirmed	2:30 am IST	Emergency Chatbot Maintenance	EU44	Einstein Bots	Emergency Maintenance
TODAY - APRIL (1)							▲
682553	Apr 21	Confirmed	2:30 am IST	Org Migrations	EU44	Core Service	Scheduled Maintenance
MAY (1)							▲
678839	May 21	Confirmed	5:00 am IST	Chat / Omni-Channel Summer '24 Release Preparation Notice	EU44	Salesforce Chat	Release
JUNE (7)							▲
642358	Jun 08	Confirmed	7:30 am IST	Summer '24 Major Release	EU44	Core Service	Release
673916	Jun 08	Confirmed	7:30 am IST	Financial Services Cloud - Summer '24 - Major Release	EU44	Core Service	Release
673925	Jun 08	Confirmed	7:30 am IST	Health Cloud - Summer '24 - Major Release	EU44	Core Service	Release

Figure 2.10: Salesforce Trust Maintenance

As you can see in the preceding screenshot, you are able to discover when Salesforce had the maintenance or when will be the next one, so you can consider this in your development plans. Some of the maintenance is also related to new features implementation, so it's worth checking them out before Salesforce releases.

> **Tip**
>
> Ever wonder when your Salesforce org will get new features? You can check it out at `https://status.salesforce.com/instances`. Just navigate to the **Maintenance** tab and search for `release` in the search bar. You may search there for major release names and their dates.

- **Version:** The current Salesforce version on which your Salesforce instance (so, also org) operates; for example, **Summer '23 Patch 18.8** (but if you are reading this book in the year 2077, it may be written Summer '77 there).

- **Region:** The region where your instance data is located; for example, EMEA

- **Maintenances Window** data: The time when Salesforce is planning to do some work related to the instance, such as implementing new features or doing some patches. There may be some access interruptions during this time:

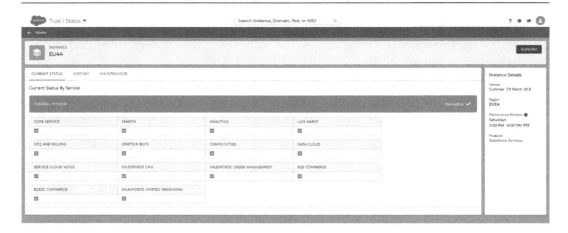

Figure 2.11: Salesforce Trust maintenance status

Tip

From time to time, Salesforce may go a bit crazy and stop working for a few minutes. This occurs very, very rarely, but of course, it can happen. If this occurs, do not panic, and refrain from blaming yourself for any potential mistake. Instead, visit the `https://status.salesforce.com/` page to check if there are any current issues or maintenance activities related to the instance where your Salesforce org is hosted.

Summary

Congratulations! You have just come through the second chapter of this book! I hope you stay motivated and ready for the next chapter. For now, let's summarize the chapter that you have just finished.

We explored the concept of multi-tenancy, where a single software instance served multiple clients. We examined the challenges faced by Salesforce in maintaining boundaries and making updates. This chapter shed light on why Salesforce employed a multi-tenant architecture. Thanks to comparison to a hotel where guests shared common resources but had isolated, private rooms, you should now remember better what multi-tenant means.

Next, we unraveled the notion of Salesforce orgs, their origins, and the various types in use. We learned about the lifecycle of an org and received practical best practices to address user and administrative concerns effectively.

In the final section of this chapter, we introduced Salesforce PODs, which are essentially server locations for Salesforce data. These instances can host multiple distinct Salesforce orgs, each with its own security measures, ensuring data isolation and protection. This chapter provided a solid foundation for understanding Salesforce's architecture and organization.

In the next chapter, you will learn about the Salesforce data model and data management.

3

Getting to Know Data Management

The world is constantly evolving, with technology advancing a few steps each day, and in this world, data reigns. Thanks to the data, we are able to find the right pattern for further steps. However, this pattern must function within a certain structure, and every system is a structure. As a result, we are able to get used to it, learn it, and eventually start using it. We can compare this to a city, which has its infrastructure, buildings, residents, and much more. When all these elements are in place, the system functions efficiently. It becomes clear who holds responsibility for various infrastructures such as the sewage system, electrical network, and water access. This allows us to verify who can help us in case of problems. As I mentioned, it is like the Salesforce system; it has its own objects, applications, records, and fields. Records are created or imported, and when they are no longer needed, they are deleted.

Data is more than just numbers or text; it is the driving force behind the decision-making process in an organization. As a leading brand in CRM systems, Salesforce understands the power that lies in data and knows how to utilize it, offering a plethora of tools for managing it in the process. However, before diving into the finer details of the system, it's crucial to understand the basics of Salesforce. As Terry Pratchett once wrote in his book *Going Postal*, *"Well, after all, one has to learn to walk, sir, before a person tries to run."* Therefore, in this chapter, you will find topics such as the following:

- Overview of the data model
- Salesforce apps
- Objects, fields, and records
- Importing and exporting data
- Deleting data
- Record types
- Custom settings and custom metadata

Overview of the data model

We step into our digital world of Salesforce. The digital realm is based on logic, on zeros and ones, ensuring that an action will trigger a reaction (unless something goes awry with the Flow, and we receive a charming red error message). It is a bit like digital karma; every action will return to you. Through our actions in the system, we can adjust the rules of digital karma. However, the rules work on certain conditions, and to meet these conditions, structure is important.

The structure of Salesforce is fairly simple. It is based on the main elements, which are the skeleton of this system, with additional elements attached to it.

So, what comprises the main elements? Let us start with objects.

On Cloudy Street, there stands a house. This house is quite ordinary – single-story, built from old (soulful) bricks, and on the door of the house, the name Salesforce is written. If we enter this house, the first thing that catches our eye will be the furniture, which adds warmth and the impression that someone lives in the house.

The same is true with objects; they furnish our CRM, and thanks to them we can enter accounts and contacts. And just like with items of furniture, there are those that are standard, such as a sofa or dining table, and those that are custom, such as a hand-shaped armchair or a bathtub that looks like a pirate ship. Standard objects are given right upon entering a fresh Salesforce, and custom objects, well, only imagination limits us in what we want to create. But what would happen if these objects lived without interactions among themselves – for example, the sofa would only be for sitting and staring at the wall, and you couldn't watch Netflix or read your favorite book on it? Objects need to have relations among themselves. We connect them in business chains, and although, for instance, there is no connection between Case and Opportunity, we can easily create one. It will enable attaching the opportunity of a sold product to a complaint. Salesforce has provided the opportunity for objects to "date" each other. The idea behind this is connecting objects to create something much larger.

The object is characterized by something, primarily its name and the fields that are located on it. This is the unique DNA of every object. These fields define what data the object is supposed to collect. They can be modified by administrators in any way they see fit. Records in the system are created in the likeness of the original – in other words, the recorded fields on the page layout of a given object. Is that all? Of course not (sorry!). Everyone is entitled to a bit of privacy, records included. By using the appropriate settings, we are able to set user access to specific records. Good filtering will set visibility only on desired records. Salesforce is an exceptionally open system (in this safe manner), allowing the system to connect with external tools, (i.e., integrations). With the right API settings, Salesforce could even brew us coffee in the coffee maker if only it had internet access and an open API. And who watches over all this? We do, the Administrators, Consultants, and Developers. But what kind of care would it be if it weren't of the highest quality? And here is where *best practices* come in; thanks to them, you know what to do and in what order. Salesforce shares its best practices at every step, in trailheads, knowledge articles, and other aids, and importantly, we also share this knowledge with you in this book. Use these aids every time you do not know how to carry out certain changes.

But the structure needs to be available somewhere and needs to be accessible. That is why Salesforce brings us applications. We'll learn more about these in the next section.

Salesforce apps

Have you ever wondered how you can facilitate users' access to certain functions in the system? So that they do not scratch their heads wondering where the opportunities are located, or so they do not have to leaf through every item in the App Launcher looking for cases? That is what apps are for. But remember, these are not the kind of apps you have on your phone, such as Candy Crush or an app for cute photos; no, these are collections of specific objects.

Imagine that your colleagues from the accounting and service departments come to you, each of them with different requirements and completely different needs. Now imagine that Mr. John from the service department has to search for a screwdriver among the papers on Ms. Jane's desk, and Ms. Jane has to find a calculator in Mr. John's toolbox. However, through these searches, they may fall in love with each other.

So, what is an app? An app is a collection of objects, tabs, reports, and charts, which create dedicated tools for specific needs. So, in the Sales app, you will not find Cases, and in Service, you will not find Price books or Products.

> **Tip**
>
> If you want to create your own custom application, go to where it all starts, which is **Setup**. Then, in **Quick Find**, find **App Manager**. There, you will be able to see **OOTB** (**out-of-the-box**) apps and those created by someone else. Each app has information displayed such as **App Name**, **Developer Name**, **Description**, **Last Modified Date**, **App Type**, and **Visible in Lightning Experience**.

While creating applications in Salesforce, you are able to customize them very well. When creating a new application, you have two types to choose from. However, it is the New Lightning App that we build step by step, and the other is the New Connected App:

- **New Connected App** is a framework that allows us to connect Salesforce with external applications using API. This enables access to external tools within Salesforce. The most crucial elements in Connected Apps are authentication, ensuring security during the connection, and access management, assigning specific permissions.

- **New Lightning App** is an application built more often than the first one. Building this application does not require access to external APIs, making its construction take place only within our org.

A new application can be very well tailored to the needs of the users. Here are some tools that we use during creation:

- **App details and branding**: This is essentially our foundation. Here, we provide the name of our application, the developer name, the description (remember to always add it), the image that will appear as the branding of our application, and the color matching our aesthetics (sometimes it is extremely important – bright colors are not recommended).

- **App options**: These settings allow us to grant permissions for display on the phone, computer, or both at the same time. The setup we will use (Normal or Service Cloud), navigation style (there are two, standard or console – this can be found in the Service app), and application personalization – in other words, disable end-use personalization of nav items in this app (whether the user can add and remove objects from the application) and disable temporary tabs for items outside of this app (i.e., adding objects from other applications temporarily).

- **Utility items (desktop only)**: In this option, you can add additional application extensions such as CTI telephony, notes, macros, and many others. They are displayed at the bottom of the page. When clicked, they open, allowing the use of new functions. An example provided by Salesforce can be seen in the following figure:

Figure 3.1: Utility items

- **Navigation items**: Here, we add the applications that will be available in this application. With available items and selected items, we can select several applications and use the Add arrow icon to move them or remove them with the **Remove** arrow:

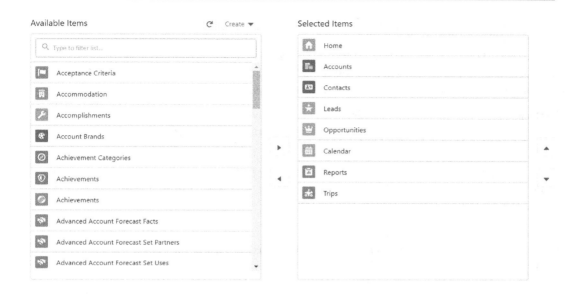

Figure 3.2: Navigation items

- **User profiles**: And here, we assign access permissions. We can add application access for a specific user profile.

You can customize your applications with these many options. And with that, you have learned about applications in Salesforce, or collections of objects. You now know how to create a new application, and how to add the appropriate objects and additional options to it. Also, you know how to make it available to the appropriate users. To summarize, applications are collections of objects, and since we are talking about them, it is objects, fields, and records that the next section will be about.

Objects, fields, and records

Just as the body consists of cells, Salesforce is also a collection of various elements. Some are larger, others significantly smaller, but each of them holds great significance. In this section, we will discuss objects, fields, and records. With these components, an administrator can create a unique tool with all the conveniences. Let us begin.

Objects in Salesforce

Objects in Salesforce are fundamental elements, without which Salesforce could not exist. For how could you find, for example, Contacts among thousands of records of one kind that are not grouped in any way? It is a bit like buying chips – with no specified flavor, just chips, and hoping to pick out the paprika-flavored ones. Would there be a guarantee you would pick it out the first time? Not necessarily. That is why Salesforce designed its structure resembling its prototype, Excel.

I think I will not be lying if I say that everyone knows at least a bit of Excel and can use it (even my eight-year-old daughter Zosia loves coloring cells). Indeed, there will be people who can use functions, macros, and other advanced tools in an incredible way, and others will simply enter a guest list for their upcoming wedding. Why am I talking about Excel? Each object can be compared to a separate spreadsheet, meaning it is a separate collection of elements with the same characteristics.

Thanks to this, we know that in the collection of people invited to a wedding, there won't be a list of gifts and a budget for the upcoming year (although nothing stops you from creating such a digital Frankenstein's monster).

Objects in Salesforce are precisely such separate collections. In the OOTB version, you can find default objects. Among them will be celebrities such as Account, Contact, Lead, and Opportunity, or the lesser-known ones such as Contract, Product, and Case.

How do we categorize objects? Well, we can divide objects into two types – standard and custom. Simple? Certainly. Standard objects are the ones I mentioned previously. So, upon opening a fresh org (yes, you now know the meaning of this word), you will be able to create a new Account, add contacts to it, and create opportunities with them. But what if, to our contacts, we want to add a customer preference tag, which will allow us to easily find all customers who like watermelon for breakfast (everyone wants to find such customers, right?)? Then we need to create custom objects. Such objects will contain data that Salesforce didn't think of before. To not leave you without examples, here they are:

- **Projects**: An object that allows us to manage projects. Attach users to it and store their deadlines and all information related to the projects.

- **Employee certifications**: This object will hold data such as the expiration date of the certificate, details of the user holding the certificate, and its description. Thanks to the expiration date of the certificate, we can build automation to remind us about the renewal, 10 days before the upcoming deadline. Useful? Absolutely!

- **Technical documentation**: In the OOTB system, we won't find an object where we could provide complicated technical details. If your company needs to store such data, create this object for them and add the appropriate fields. They will be thrilled.

These three details are just the tip of the iceberg. The possibilities are vast; the only thing that limits us is our imagination. To sum up, custom objects meet the specific needs of a company. Therefore, during the course of a consulting project, a consultant, solution architect or administrator designs new objects and makes them into a necessary structure for the company.

Objects themselves are not only a dedicated element for certain needs, but they also hold the necessary data. By creating fields or relationships between objects, administrators are able to create data structures that reflect business needs.

> **Tip**
>
> When creating new objects, always write the object description. This way, a new employee or you, even after a long time, will know why such an object was created.

Fields in Salesforce

And so, we've reached fields. Things get very interesting here.

What are fields? Well, going back to the Excel example, if we were to use it, fields are the headers of columns. Thanks to them, we know what data should be found in the column, and whether it should be text or maybe a number. The name on the header can tell us a lot.

So, fields in objects allow us to store data in objects. At present, there are 23 types of fields, which we will discuss later in this chapter. Fields have very important functions in the system, of course, the most important of which is the previously mentioned data storage. The others are as follows:

- **Data validation**: Fields can have defined data entry rules, ensuring that the database will contain correct data.

- **Data classification**: Fields allow for the segregation and classification of data. This facilitates data searching.

- **Relationships between objects**: Thanks to the **Lookup type** field, you can build a relationship between one object and another.

As a curiosity, I want to pass on some secret knowledge about limits. Everything has its limits – OK, I'm not sure about space, so in that case, almost everything has its limits. The same goes for objects. Well, objects have a limit of 500 fields, 40 custom relationship fields, and 25 rollup summary fields. It's important to remember these limits. And if you're close to the 500th field, remember that the performance of the object may be unsatisfactory. Maybe it's just better to not do it.

When creating fields, there are a few elements that we cannot forget. First and foremost, the name. This is the most important element of the field; without it, it can't be saved. But to not make it too easy, there are two fields responsible for the name – the **Field Label** field, which is the name visible on the frontend, and **Field Name**, also known as the API name. It is worth applying a naming convention to it, thanks to which we can build fully functional logical groups. To make it more understandable, I'll give you an example. Let's say we have a Contact object with two record types: employees and clients. On employees, we have a field containing employee insurance data. And on clients, we have a field containing car insurance. But both fields have the Field Label – Insurance. Using best practices, it would look exactly like this:

- **Object: Contact, Record type: Employee, Field label: Insurance, Field Name: Employee_Insurance__c**

- **Object: Contact, Record type: Client, Field label: Insurance, Field Name: Client_Insurance__c**

With this solution, when exporting data, you will easily find the field you are interested in.

As I mentioned earlier, in Salesforce, we have as many as 23 types of fields to choose from. Types are dedicated to specific needs. Let's list them all:

- **Auto-Number** is a field used for automatically generating a number with a pre-designed structure. The number is automatically incremented for each new record – for example, `ThisBookIsGreat-0001`, `ThisBookIsGreat-0002`, and so on.

- **Formula** is a special field. It's non-editable and performs specific calculations or operations on the values of other fields. It has applications such as calculations, conditional logic, text manipulation, date/time formatting, and so on. For example, if, in the Opportunity object, there is $10k in the Amount field, the Formula field should show Hot Deal. The following shows how to break it down:

```
IF(
Amount >= 10000,
"Hot Deal",
"Not Hot Deal")
```

Remember to select the "Text" formula field type.

- A **roll-up summary** is a read-only field that displays the sum, maximum, or minimum of fields on records in a relationship. A roll-up summary doesn't work with lookup fields, only **Master-Detail**. For example, every "Gift" object record has information about the number of presents. On the `Contact` object, which is in a relationship, all the gifts sent to a particular customer are counted.

- A **lookup relationship** is a field that establishes a relationship between objects. If we need to connect something, this field will help us. By clicking on it, we can find any record from the previously selected object. For example, the `Contact` object is connected to the `Plan` object. This allows us to assign a specific plan to a specific customer.

- A **master-detail relationship** is a field somewhat similar to the **Lookup** field, but also a bit different. Sounds mysterious, right? Here, we connect two objects; one will be the parent object, and the other the child object. It's a one-to-many connection. This field allows the creation of a **Roll-Up Summary** field on the parent object – for example, the **Order** object and the **Order Item** object. One order can have several products.

- An **external lookup relationship** allows for connections outside the org. Yes, it's true, you can establish a connection with another org (linked by Salesforce Connect, a tool for connecting two separate orgs) using an external ID – for example, two orgs, one for-profit and another non-profit. In the profit org, they want to know whether customers are making donations. External lookup allows for connecting them in the profit system.

- A **checkbox** is one of the most inconspicuous fields but has great power. On the page layout, it appears as a small square with a check or uncheck. It only has two values: TRUE and FALSE. This field can be used as a validation field for automation (in this case, it can be hidden from users' view) or as a marker for a particular record. Example: When creating an Opportunity record - Subscription, the Subscriber checkbox is checked by automation. This way, the client is added to the mailing list to receive newsletters.

- **Currency** is a field used to store values in the system currency – for example, Purchase values on Opportunity.

- **Date** and **Date/Time** are two fields that allow for entering a date or a date with a time. Example: Entering the Policy End Date on the Insurance object.

- **Email** is a field validating the proper structure of the entered text. The field checks whether the entered text is an email address – for example, Personal Email on the Contact object.

- **Geolocation** is a field holding latitude and longitude data – for example, Treasure location on the Map object.

- **Number** is just a number – for example, the number of products ordered.

- **Percent** means percentage value – for example, a discount rate on Opportunity.

- **Phone** is where you enter a phone number. This field formats the number appropriately – for example, personal phone on the **Contact** object.

- **Picklist** and **Picklist (Multi-Select)** are fields with the option to select one or more from a list of values – for example, **Contact** object, **Candy type** preferences. Chocolate only? Or maybe chocolate and gummies?

- **Text, Text Area**, **Text Area (Long)**, **Text Area (Rich)**, and **Text (Encrypted)** are text fields with various properties. Let's start with the last one (did I surprise you?), which only works in Classic. So, if you use the Classic interface, be my guest and set up a field that allows for storing data in encrypted form. Text (up to 255 digits), Text Area (up to 255 digits with separate lines), and Text Area (Long) (up to 131,072 digits) are text fields where we enter text without any rocket science. But Text Area (Rich) allows for a bit more, such as using text formatting. We can use the most popular formats such as bold, italic, and many more. Example: Product description.

- **Time** allows for entering the local time – for example, Customer time.

- **URL** is the place where you can put any URL – for example, a client website.

Wow, that was quite a lot. I must admit that I have never considered how many types of fields Salesforce offers. If you were to ask me whether you must memorize all of this, I would say it is worth it, but you will get used to it as you start working regularly with Salesforce, as creating new fields is one of the many tasks you will perform or are already performing as an administrator.

Once you choose the type of field, give it a name, and assign the appropriate values, you must decide whom you want to grant access to. You will see a group of all profiles and can choose for whom this field will be visible and for whom it will be read-only. The next step is to add the field to the appropriate page layout. And here, we have two options: to add or not to add, yes, I know, surprising. If you add the field, it will appear in the first section of the page layout, but possibly not where you would like. If not, open the page layout and find your new field, which you can place anywhere.

Another essential feature is default values. Almost every type of field can be assigned a default value. Every time you create a record, such a value will be entered for the newly created record. For instance, let us say you run a podcast on Spotify, and you collect listener data in the system. You have a premium program where, once a listener listens to at least 50% of the material from your channel, you send them a shirt. By default, a Contact is not a Super Fan, so the default value is just Fan. So, when creating a new contact, the Title field has the value Fan. But when all the conditions are met, the field is overwritten with the new value Super Fan.

Access can practically be granted or revoked to any element in the system, and the same goes for fields. We can also assign specific properties to fields. We have two such options: **Read-only** and **Required**. **Read-only** is assigned to fields that cannot be edited. The value might be overwritten during creation, integration, or automation. **Required** is an option for those who forget things. How often did you check something in the system and lack a phone number or email address? That is where the required option comes in handy. Thanks to this feature, the system will enforce the entry of a value in the field required.

Records in Salesforce

And we have arrived at records. I will go back to the Excel example. We have already got a spreadsheet as an object, and columns as fields, but we are missing rows. Records are our rows. They are the basic units of data in our beloved system. They represent customers, accounts, opportunities, or leads. Fields that store data about that record make up a record. But how do you differentiate two records with the same data? The solution is simple: the Salesforce ID. The Salesforce ID (visible after export or in the URL bar upon entering a record) consists of a string of characters, either 15 characters (case-sensitive) or 18, which is the case-insensitive version, making it easier to cooperate with external systems.

Creating records takes place on a specific object. Upon entering an object, you will easily find the **New** button, which will activate a popup with empty fields to fill in. If you want to edit a record, that is tough – just kidding. Upon entering the list of records of a given object, you will find a drop-down arrow on the right side of each record, which, when clicked, will display the edit option.

And this way, we have reached the end of this section. You have learned what an object, field, and record are. What types of objects can you find in the system, and how many types of fields can you see when creating a new one? It is crucial to feel comfortable with this list so that you know which field to propose when someone wants to add percentages and date/time in the system.

Now, I invite you to the next section, where you will learn the magic of data import and export.

Importing and exporting data

In this section of the book, we will learn how to export and import data to Salesforce and recognize the tools and features that support these activities. As you know, all software revolves around data. Of course, features are also important, but ultimately, data and records are the keys to success for companies using Salesforce. Therefore, mastering how to deal with data is crucial for every Salesforce Administrator, as *"Data is like garbage. You'd better know what you are going to do with it before you collect it."*

Exporting and importing Salesforce data

Before we delve into importing data into Salesforce, we'll address the export options. Often, the data you import into Salesforce is data you've previously exported and updated with additional information. For this reason, let's start by learning how to export data. Later, we will proceed to learn how to import it. There are several options available for exporting or importing data from Salesforce. Let's list them as follows:

- Salesforce Reports – Data export capabilities
- Import Wizard – Data import capabilities
- Data Loader – Data import and export capabilities
- Workbench – Data import and export capabilities

Let's now see how you can use each of these features.

Exporting data via Report Export

The full description of how to use Salesforce's reporting features will be given in the *Reports and Dashboards* chapter. Here, we will focus only on the exporting possibilities of the Salesforce **Report Builder**.

To export Salesforce data via Report Builder, let's take the following steps:

1. **Run** or **Save and Run** a report.
2. Click the **Export** button.
3. Choose **Export View**.

   - **Formatted Report** – Export the data that is visible in Report Builder. Besides the data, the groupings and report header are also exported.

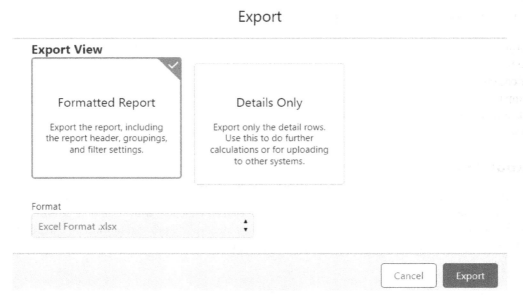

Figure 3.3: Export view

- **Details Only** – Only raw data are exported. When choosing this option, you will also be able to choose the file format and encoding. Besides the `.xls` or `.xlsx` Excel formats, you can choose the `.csv` format. You are also served with many data encoding options. Encoding is responsible for displaying proper letters/characters.

Figure 3.4: Export format

4. Click the **Export** button. The file will be downloaded and saved on your computer. If there is a large amount of data, it may take slightly longer.

Tips

I seldom use the **Formatted Report** option because the exported data can be troublesome to edit. Columns or rows may be merged, making it hard to manipulate the data in Excel or Excel-like tools. Most of the time, you'll want to use the **Details Only** export option since it provides the data in an easy-to-manipulate/update format. The .csv export format should be used when you know that you will need to import the data back to Salesforce as Salesforce importing tools use the .csv data format.

When I'm exporting data, I mostly use UTF-8 encoding as this gives me a greater chance that the letters/characters will be displayed properly when editing the data in Excel or an Excel-like program.

Import data via Data Import Wizard

The **Data Import Wizard** is a straightforward tool that facilitates importing data into Salesforce. It's available in your **Organization Setup**, allowing for online execution. You can insert both standard object data and custom data using this wizard. The Data Import Wizard is user-friendly. If you want to give it a try, simply follow these instructions to import some data into Salesforce:

1. Access the Data Import Wizard. Open the Wizard from **Setup** by entering Data Import Wizard in the **Quick Find** box and selecting the corresponding option.

2. Select the data for import. For importing Accounts, Contacts, Leads, Solutions, Person Accounts, or Articles, choose **Standard Objects**. For custom objects, select **Custom Objects**.

3. Define whether to add new records to Salesforce, update existing records, or perform both actions simultaneously:

Figure 3.5: Data Import Wizard – object choice

Specify matching and other criteria as needed, utilizing the provided question mark tooltips for more information:

Figure 3.6: Data Import Wizard – type of action

4. Indicate whether to trigger workflow rules and processes when the imported records meet the defined criteria. Here, you are just deciding whether those automations should trigger during the data insertion:

Figure 3.7: Data Import Wizard – workflow rules and processes options

5. Identify and select the file containing your data. Specify your data file either by dragging the CSV file to the upload area or by clicking on the appropriate CSV category and navigating to the file.

6. Choose a character encoding method for your file. Typically, the character encoding remains unchanged.

7. Choose either a comma or tab as the value separator.

8. Map your data fields to Salesforce data fields. The Data Import Wizard automatically maps as many data fields as possible to standard Salesforce data fields. However, manual mapping is necessary for any fields that the wizard cannot map. Unmapped fields will not be imported into Salesforce.

9. Review and initiate your import.

10. Monitor the import status. You will see when the import is imported.

> **Tips**
> Many people start their Salesforce journey using the Data Import Wizard. The reason is that the tool is kind of built into the Salesforce org's page, making it easy to access and relatively user-friendly. However, this marks the end of this romantic story. Beginners soon discover that the Wizard is not a powerful sorcerer, and that the true power of data migration lies in different tools. Fortunately, we have described all of them for you in this chapter.

Exporting and importing data via Data Loader

Before utilizing the Salesforce **Data Loader**, you must install this tool on your computer. Yes, you read that correctly! Salesforce, the platform with the cloud as its logo, pushes you to install it on your computer instead of providing this solution in the cloud. Deal with it; Data Loader is not a cloud-based tool and is not accessible via a web page. You need to act like an old-school kid, and download and install the tool on your computer or laptop. But don't worry. If you have the authorization to install software on the computer you are using, the entire installation process is quite easy and fast.

It should take around 10 minutes, and Salesforce provides you with a tutorial, detailing how this should be done: `https://developer.salesforce.com/docs/atlas.en-us.dataLoader.meta/dataLoader/loader_install_general.htm`

After the installation is completed, please follow these instructions on how to export data from Salesforce:

1. Run the Data Loader app – just run the app on your computer.
2. Log in to Salesforce and access the environment from which you would like to export data.
3. Choose one of the export options:

    *   **Export** – You can choose which data will be exported.
    *   **Export All** – All data will be exported.

Figure 3.8: Data Loader – action choice

4. Export the data.

> **Tips**
>
> If you really don't want to install anything on your computer, or if you're unable to do so, or perhaps you're using a random computer on a random internet in the middle of Bangkok (of course, this is not recommended for accessing important, sensitive data), you may want to use the **dataloader.io** feature. dataloader.io is similar to the Salesforce Data Loader but is a cloud-based (accessible via a web page) data integration platform. dataloader.io enables users to import, export, and delete data in various cloud applications, including Salesforce. I strongly recommend getting to know this tool as it has some nice features that are not available in the Data Loader desktop app.

Now, if you would like to know how to insert the data with Data Loader, just follow these instructions:

1. Open Data Loader and choose the **Insert** option.
2. Display all Salesforce objects.
3. Opt for the object that you want to use to insert your data – for example, Account.
4. Locate and select your CSV file – the file with the data.
5. Click on **Create** or **Edit a Map**, then choose **Auto-Match fields to columns**. Now, this is an important step as here, you will match the data from your file with the Salesforce object fields.
6. Proceed by clicking **OK, Next and Finish**.

Exporting and importing data via Workbench

Workbench stands as a potent collection of web-based utilities created for administrators and developers to engage with Salesforce.com organizations using the Force.com APIs. It encompasses strong backing for various Force.com APIs such as Partner, Bulk, Rest, Streaming, Metadata, and Apex. These APIs empower users to articulate, inquire, manage, and transfer both data and metadata within Salesforce.com organizations, all conveniently accessible through a straightforward and user-friendly interface within their web browser. Just to summarize: Workbench is a free but not official Salesforce tool that can be used in many situations while working with Salesforce data and metadata.

To use Workbench to export Salesforce data, do the following:

1. Sign in to Workbench – just go to `https://workbench.developerforce.com/`.

2. Place your cursor over the Queries menu item and opt for a SOQL query.

3. Pick **Bulk CSV** from the **View as** radio input field.

4. Compose your query within the **Enter or modify a SOQL query below** input field. To do this, you may select the object and its fields or just write the SOQL query from scratch:

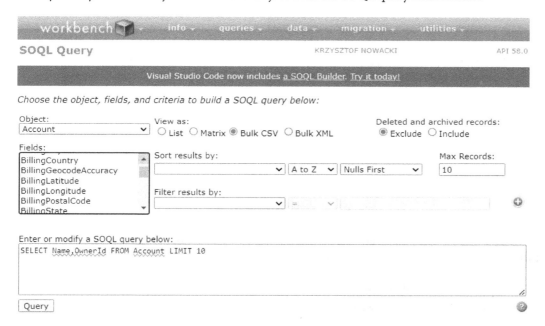

Figure 3.9: Workbench – main screen

5. Click the **Query** button. You will be directed to the **Bulk API Job** screen (if the SOQL query is error-free).

6. Be patient and wait for the **Status** to transition to **Completed**.

7. Select the download icon to download data (on the left in the **Batches** section).

Figure 3.10: Workbench – job status

OK, we have just exported data via Workbench. But do we use this tool to import it into Salesforce? Is this even possible? Yes, it is! Just follow the steps:

1. Return to Workbench and choose **data | Insert** from the top menu.

2. Choose **Object Type** and opt for the **From File** radio option:

Figure 3.11: Workbench – action choice

3. Select the file with the data that you want to import by clicking the **Browse** button and then proceed to click **Next**.

4. Click the **Map Fields** button, and then confirm the insert by hitting the **Confirm Insert** button.

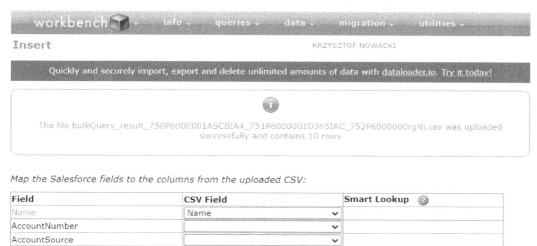

Figure 3.12: Workbench – fields mapping

5. Finally, click the **Download Full Results** button.

As we've described a few possible options to export data from Salesforce, you may now wonder which tool to use for exporting your data. The answer is simple: use the tool that best suits your situation. If you just want to export some Salesforce data, for example, to use in a presentation, the Report export would be fine. However, if you frequently work with data imports and exports, Data Loader sounds like a better fit. If you're a Salesforce enthusiast, you may also want to try using Workbench.

What's crucial to note is that each tool has its own limitations. For example, when using the Data Import Wizard, you can import a maximum of 50,000 records at a time, and you won't be able to import some important object records such as Opportunities or Cases. To bypass these limitations, you can use the Data Loader tool, as Data Loader is supported for loads of up to 5 million records and can handle multiple Salesforce objects. Understanding these differences is important as they often come up as questions on the Salesforce Certified Administrator exam.

Many times, data exporting is related to archiving the old Salesforce data. After moving it to a safe place for the future (this can be Salesforce Big Object or an external database), you may have to delete it from Salesforce. Let's cover this topic in the next section.

Deleting data

If you have the possibility to delete all Salesforce data, that means that you are someone important, maybe a Salesforce Admin. If you are not, then someone did not spot something and gave you too much power, because as the well-known saying goes: "*With great power comes great responsibility.*"

In this section, we will detail the following few ways in which data can be deleted from Salesforce:

- Deleting records manually
- Mass deleting records via Data Loader
- Flow Delete action

No worries; we will also see how to rescue the deleted data.

Deleting records manually

As simple as it may seem, you may have the right to delete records manually, one by one. It sounds like an obvious feature; if someone can create records, they should also be able to delete them. However, in the Salesforce world, to be able to delete records on some objects, the user needs to have the proper permissions. Permissions come separately for each object and can be assigned to the user, for example, via Permission Sets. The object permissions that can be used to permit users to delete records are as follows:

- **Delete** – The permission that permits users to delete records on a specific object.
- **Modify All Data** – God mode on the level of Salesforce objects. The user with this permission can edit and delete all records on a specific object.

Mass deleting records via Data Loader

Imagine you need to delete 10 records from Salesforce. Of course, you can do this one by one. Easy-peasy. But try doing the same with 100 or 1,000 records. Yeah, it's still easy, but incredibly time-consuming. Now, let's imagine you'd prefer to enjoy a good coffee break in the office instead of clicking the delete button in Salesforce 1,000 times. Not so hard to imagine, right?

If you'd like to delete records more efficiently, then you need to perform this as a mass action. This can be achieved by using the good old desktop-only Data Loader.

To delete records via Data Loader, you need to follow the following steps:

1. Export the Salesforce record data that you need to delete. You can use Report exports or Data Loader to export this data. What is important is the Salesforce record ID needs to be included in the export data.

2. Use the delete feature in Data Loader. It works similarly to importing, which was described in the previous part of this chapter, but instead of inserting data that should be imported or updated, you will use the report of records that need to be deleted. That's why you needed the record IDs when you were exporting the data. Based on the IDs, Salesforce will know which records need to be deleted.

Deleting records via Salesforce Flow

In the Salesforce world, there are or have been many automations that can create or update records, but not many that can delete them. Besides coding, there is a low-code solution that can be used to delete records. Moreover, it can be used to delete both single and multiple records. Ladies and gentlemen, let's once again welcome Salesforce Flow. Salesforce Flow is a powerful tool that, in addition to creating and editing records, offers the option to delete them. When creating custom Flow automation, you can use the **Delete** action to perform record deletion. This action can be triggered after the creation, updating, or even deletion of other records. What's more, you can choose to run Flow in system mode. For example, even if the user who triggered the Flow automation doesn't have permission to delete some records, they will be deleted as the system mode will bypass this restriction. Powerful? Yes, it is! Will you use it? Yes, you will!

Restoring the data

Sometimes life can be tough, requiring you to save an entire company, some specific co-workers, or at least one person, but most importantly, yourself. This can happen when someone (was it you?) "accidentally" deletes some data from Salesforce. Besides answering the question of how this happened, the first thing you would need to do is save the day and restore this data. Thankfully, this is not hard to do, as Salesforce provides us with a Recycle Bin feature. It's accessible as a tab and is really easy to use.

To find proper data, you can use two predefined List Views:

- **My Recycle Bin**
- **Org Recycle Bin**

Figure 3.13: Recycle bin in Salesforce

To restore the data, just mark the specific record and restore it, or mark a few records and do this as a mass action. As simple as that. What is important is that currently, there is no "restore all" option in Recycle Bin, so for this reason, Salesforce recommends using the Workbench tool to perform this action.

Now you know how to restore deleted data. You can now say "I'm the king of my trash can!" as Oscar the Grouch from Sesame Street said.

> **Tips**
> Restoring data is quite easy but you need to remember that the deleted data will not be in the Recycle Bin for long. After deletion, the record "sits" in the Recycle Bin for 15 days. After this time, the records will be purged from the Recycle Bin and can't be recovered.

You've just gained valuable knowledge about deleting and recovering records in Salesforce. What I can whimsically wish for you is that in your Salesforce journey, you won't have to recover records unintentionally deleted by users. However, if you ever need to, you already know where to find them. Although, remember that regardless, it's worth considering how to regularly back up your data in Salesforce. We won't delve into this topic in this book, but there are plenty of options to do so. I encourage you to explore this topic on your own.

To conclude the topic of imports, exports, data deletion, and recovery, I invite you to another topic in this book. In the next section, we will discuss the interesting subject of Salesforce record types.

Record types

Besides Salesforce security topics, record types is a concept that, if not described properly, is hard to understand for Salesforce knowledge seekers. So, what are Salesforce record types? In simple words, record types represent different types of records. Let's consider a simple example: there are multiple companies that interact with the company you are working for, such as prospects, customers, competitors, and partners. In Salesforce, all these types of companies can be stored as Account records. To differentiate the company types, you can use record types. Just create the following record types: Prospect, Customer, Competitor, and Partner. Then, assign the proper Account records to their respective types.

Of course, the record type feature is more complicated than shown in the previous example. But do not worry, I will try to explain it further. Please see the following for some information about record types that may help you understand what this feature is all about:

- Record types are not mandatory in Salesforce. If there is no need to create record types, you don't have to. For example, the types of records (such as Customer, Partner, and Competitor) could be marked with picklist field values instead of record types. However, it's important to know that using record types offers additional features that can enhance the visibility of this record separation in the database and ultimately on the user interface.

- Different record types can have different page layouts and user interfaces. Thanks to this, you can display different user interfaces (fields, sections, and related lists) based on the record's record type. For instance, Customer Accounts can have different sets of visible fields compared to Competitor Accounts. Achieving this is not as straightforward when using just picklist values (a similar but not identical effect could be achieved when using Dynamic Forms and field display filters).

- The use of record types can be assigned to users, making them the default for some and allowing a choice when creating records. For example, when creating an Account record, Salesforce will prompt the user to specify whether it's a Prospect, Customer, or Competitor record.

- Record type user assignment is stored in Profiles and Permission Sets. You can set which record type is the default for a user. Additionally, you can decide which record types could be used by a user while creating records on specific Salesforce objects, such as Accounts or Opportunities. For instance, some users could create Prospect or Customer records, while others could create Competitor records.

- Record types related to Opportunities or Cases need to be supported by previously created sales processes or support processes.

- The record type of the record is set when the record is created (if it is set up in the company), but it can be changed if needed, even after the record is already saved. To do this, you need to use a dedicated action button.

- On the record's user interface, the record type name is displayed similarly to Salesforce fields. It can be shown on the record user interface, header, list views, reports, and so on.

- Record type values have their own names and IDs, which is important to know when creating Salesforce automations related to records that are associated with record types.

> **Tips**
>
> When creating Salesforce automations achieved with Flows or code, try to not hardcode record type IDs. The simple reason is that if you hardcode them, the automations or code will not work on your other Salesforce environments/Sandboxes – for example, on a testing Sandbox. This will happen because during the Sandbox creation, the Salesforce will copy the production record types' names but will assign them different IDs. Instead of hardcoding IDs, try to use the record type object (yes, there is a separate Salesforce object to store record types) and use the record type's Name or Developer Name.

Record types are an important functionality that you will surely use at some point. It's important that the use of this functionality is deliberate. It won't always be necessary; sometimes, using a regular field such as a picklist will suffice.

Now that you know what record types are, it will be easier for you to make decisions, so I invite you now to learn about the next Salesforce functionalities. Let's talk about custom settings and custom metadata.

Custom settings and custom metadata

Custom settings and custom metadata are highly similar concepts within Salesforce. As you may know, Salesforce occasionally develops features that serve similar purposes, such as Process Builder and Flows, and eventually opts for a single solution. A comparable situation arises with custom settings and custom metadata – they can achieve similar objectives. So, why did Salesforce create these distinct features? The answer is straightforward: because they can! Just kidding. Custom settings were introduced prior to custom metadata, so we could say that custom metadata is a sort of evolution of custom settings. This is, of course, a significant oversimplification, so let's quickly compare these features:

	Custom settings	**Custom metadata**
Suffix	"__C", similar to a Salesforce custom object	"__mdt"
Types	List or Hierarchy	Does not support hierarchy based on Profile or User
Deployment	You have the flexibility to move your custom setting structure (metadata) using standard deployment approaches. However, it's important to note that transferring the actual data within the custom setting, whether through packages, the Metadata API, or change sets, is not supported. Sad but true: you would need to move the records manually.	Custom metadata types outshine custom settings in the sense that they offer the capability to deploy both metadata structure and data/records using change sets.
Supported field types	<ul><li>Checkbox</li><li>Currency</li><li>Date</li><li>Date/Time</li><li>Email</li><li>Number</li><li>Percent</li><li>Phone</li><li>Text</li><li>Text Area</li><li>URL</li></ul>	The same as supported by Custom Settings minus:<ul><li>Currency</li></ul>plus:<ul><li>Metadata Relationship – creates a relationship that links this custom metadata type to another metadata type, entity definition, entity particle, or field definition</li><li>Picklist</li><li>Text Area (Long)</li></ul>

	Custom settings	**Custom metadata**
Formula fields, validation rules, Flows usage	Can be used by Salesforce formula fields, validation rules, and Flows.	Can be used by Salesforce formula fields, validation rules, and Flows.
Apex usage	You can perform CUD (create, update, and delete) operations on custom settings in Apex. Not visible in test class without "SeeAllData" annotation. Custom settings data can be accessed using instance methods, which allows you to avoid making SOQL queries to the database.	You can't perform CUD (create, update, and delete) operations on custom settings in Apex. Visible in test class without "SeeAllData" annotation. You can get the custom settings with SOQL.
API usage	SOAP API	Use SOQL to access your custom metadata types and retrieve the API names of the records of those type[s].

Table 3.1: Custom settings versus custom metadata comparison

As we've discussed, there are some similarities between both features. Before we take a side and recommend the best feature, let's take a closer look at each of these features, including how to create them.

Custom settings

Custom settings resemble custom objects, allowing you to tailor organizational data. However, unlike custom objects that have records associated with them, custom settings enable the utilization of custom datasets across your organization. Additionally, custom settings permit differentiation of specific users or profiles based on customized criteria. Custom settings can be accessible and then used further to create Salesforce code, validation rules, Flows, and so on. So just to summarize this in simple words: Salesforce custom settings are a way to store configurable data that can be accessed across your Salesforce application. Let's have an example to understand this better. Let's say you have a Salesforce app used by salespeople in different regions. You could create a custom setting called "SalesRegionSettings" with fields such as "Sales Target" and "Sales Manager." Each region can have its own instance of this custom setting, allowing you to set specific sales targets and managers for each region.

Let's expand on this. There are two sales teams: North and South. These teams can have different sales targets and different sales managers:

- SalesRegionSettings for "North" could have a sales target of $100,000 and a manager named John Doe
- SalesRegionSettings for "South" could have a sales target of $80,000 and a manager named Jane Smith

Now, in your Salesforce app, you can use these custom settings values dynamically. When a salesperson from the "North" logs in, the app can read the "Sales Target" and "Sales Manager" from the "North" custom settings instance to provide region-specific information and goals for that user. It allows for easy customization and adaptability within the Salesforce platform.

I hope this was clear enough to understand custom settings. Now let's see how to create one.

Follow these steps to establish and employ custom settings:

1. Navigate to **Custom Settings** in **Setup**.
2. On the **All Custom Settings** page, choose to create a **New** custom setting, or modify an existing one by clicking on its **Label** name.
3. Fill in the required fields with the following details:

    - **Label**
    - **Object Name** – the name used when the custom setting is referenced by formula fields, Apex, or the Web Services API

4. Assess the available protection and privacy choices. This refers to the following:

    - **Visibility**:
        - **Public**
        - **Protected** – regardless of the package type (managed or unmanaged), the subsequent components are accessible: Apex, Formulas, Flows, API, provided users possess the **Customize Application** permission or have permissions granted through profiles or Permission Sets
    - **Setting Type** – visibility for custom settings can be set only in a developer, sandbox, or scratch org:
        - **List** – list setting enables access to data without profile or user dependency built in – for example, two-letter state abbreviations, international dialing prefixes, and catalog numbers for products
        - **Hierarchy** – hierarchy settings enable customization of your application based on various profiles and/or users – for example, data or feature dependencies based on Salesforce profile or user for integration purposes

- Granting permissions on Profiles or Permission Sets – when editing Profiles or Permission Sets, you can add the custom setting in the **Enabled Custom Setting Definitions Access** section

- Behavior of Apex, Visualforce, and Aura

5. **Save** your settings.

6. Incorporate fields and input data, the core of the feature. Here, you will create fields and records. It's very similar to creating object fields and records.

7. Utilize the custom setting data within your application through formula fields, validation rules, Apex, or SOAP API references.

Tips

When you were trying to create a new custom setting, you could see that some setting options were greyed out. For example, you saw maybe that there was only the possibility to create the hierarchy setting type. So, this is because Salesforce recommends using custom metadata for creating list-type metadata. You can unlock it in the Salesforce settings as described at `https://help.salesforce.com/s/articleView?id=000383145`, but we recommend not to do so and to follow Salesforce recommendations. When creating a custom setting directly in the production environment, the **Visibility** setting (**Public** vs. **Private**) is also greyed out. This is because the visibility for custom settings can be set only in a developer, sandbox, or scratch org.

Custom metadata

Custom metadata is data that can be utilized to construct apps and features within Salesforce. Instead of constructing apps directly from data, you can build apps defined and driven by their own types of metadata. The created application metadata is flexible for customization, is deployable, can be packaged, and is upgradable.

Let's consider an example. Imagine you have a Salesforce application for event management, and you want to categorize events based on types such as "Conference," "Webinar," or "Seminar." Instead of coding these event types directly into your application, you can leverage custom metadata. You would create a custom metadata type called "Event Types," including records such as "Conference," "Webinar," and "Seminar." If you introduce a new event type in the future, you simply add a new record to the "Event Types" custom metadata. Your application can then dynamically read these records and display the available event types, allowing for easy updates and additions without altering the application's code.

In essence, custom metadata serves as a dynamic configuration tool, enabling you to manage your application's settings and configurations without requiring modifications to the actual code. This enhances your application's flexibility and ease of maintenance.

I hope this was clear enough to understand custom metadata. Now, let's see how to create one.

Follow these steps to establish and employ custom metadata:

1. Navigate to **Custom Metadata Types** in **Setup**.

2. On the **All Custom Metadata Types** page, choose to create a **New Custom Metadata Type**, or modify an existing one by clicking on its **Label** name.

3. Fill in the required fields with the following details:

    *   **Label**, **Plural Label**, and **Vowel Sound** – denotes the type in a user interface page

    *   **Object Name** – exclusive name for referencing the object when using the API, which helps avoid naming conflicts

    *   **Visibility** – determines who can view the type:

        *   **Public** – accessible to all, regardless of the package type, with specific permissions such as Apex, Formulas, Flows, and API access based on permissions granted

        *   **Protected** – visible only to Apex code within the same namespace, and the name of the type and record are visible if referenced in a formula

        *   **PackageProtected** – in a second-generation managed package, visible exclusively to Apex code within the same managed package; the name of the type and record are visible if referenced in a formula

4. Save your settings.

5. Incorporate fields and input data, the core of the feature. Here, you will create fields and records. It's very similar to creating object fields and records and also similar to custom settings.

6. Utilize the custom metadata within your application through formula fields, validation rules, Apex, or API references.

As you can see, there are some similarities between custom settings and custom metadata. In the words of a popular Hollywood movie, Highlander, *"there can be only one."* We need to monitor future releases to see whether Salesforce will take a step toward merging, leaving one, or replacing these functionalities. We know that similar developments were occurring with other Salesforce features such as Workflows, Process Builder, Visual Force, Aura, and many more. Meanwhile, we recommend leaning more toward custom metadata. Even Salesforce encourages you to consider this direction, as when creating a new custom setting, there is a prompt promoting custom metadata.

Summary

Congratulations! You've reached the end of *Chapter 3*, and your knowledge of Salesforce has noticeably increased. As you've probably noticed, Salesforce has the capability to store data, modify it, and, if necessary, delete it. It has a plethora of data types and custom metadata that can be utilized in various ways. As we mentioned, Salesforce is data, and data is Salesforce; without it, it's just a good-looking system with a pretty cool logo. I believe you'll easily recognize the importance of data during your first project, where you'll configure every element, set fields, reports, and finally dashboards. And you'll look sadly at the charts that won't display anything. Using the knowledge from this chapter, you'll perform a data transfer, and then, just then, you can shout, "It's alive! It's alive!" Yes, your charts will come to life, displaying all the values, and previously set functionalities will finally work as you designed.

What makes a beautiful-looking system, besides your configuration? The Lightning experience, which we'll discuss in the next chapter. There, you'll find some information on the transition from Salesforce Classic to Lightning, a bit about Lightning components, and many other exciting features! I invite you to the next chapter, where you will learn about Salesforce Lightning.

4

Lightning Experience

In this chapter, we will discuss the Lightning Experience. And no – we are not talking about the famous AC/DC song, *Thunderstruck*; we are talking about the Salesforce interface that has won the hearts of users. In this chapter, we will discuss what Classic was and what Lightning is now. We will answer important questions about why Lightning has gained so many fans and, importantly, we will learn how to operate within it and even what tools can be used to customize it to our needs. The last element we will touch upon in this chapter is Salesforce on the phone, which saves the life of more than one salesperson hitting the road every day.

In the chapter, we will cover the following topics:

- Transitioning from Salesforce Classic
- Introducing Lightning components
- Customizing Lightning pages – Lightning versus Classic
- Record page layouts
- Related lists
- Salesforce for mobile

Transitioning from Salesforce Classic

Before you start reading this chapter, answer a very important question – have you ever had any experience with Salesforce Classic? No? You can very easily switch to this interface and see what the life of a user and administrator looked like some time ago. It's very simple; let me show you in the following screenshots how to do it:

1. In the top-right corner, find the little astro in **out of the box (OOTB)** or another icon labeled **View profile** (highlighted) and click on it:

Figure 4.1: View profile icon

2.   Then, after clicking, a new window will open with a few pieces of information, such as your user data, org name, link to **Settings** and **Log Out**, display density, **Add Username**, and **Switch to Salesforce Classic**. Don't be afraid – press it, and here begins your *Back to the Future* moment:

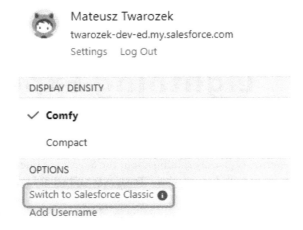

Figure 4.2: Switch to Salesforce Classic option

3.   After clicking, you will be taken to the classic version of Salesforce, which looks like this:

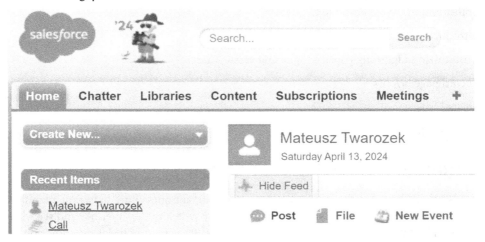

Figure 4.3: Salesforce Classic

4.   You can return to the previous version in an equally simple manner. Look at the top, and you will find a small inscription with a small lightning icon labeled **Switch to Lightning Experience**. Simple? It seems so to me.

5. When Salesforce introduced a new interface for users in 2015, it represented a kind of digital renaissance; it was a complete change for users, administrators, developers, and others associated with this system. Lightning ushered in a new era bringing improvements in terms of interface, customizing the system to user needs, and application development. Its modern look encouraged users to use it. But remember – every change has its supporters and opponents. I remember very clearly the year 2018 when I changed jobs and worked for a company that was a partner of a large consulting firm. 3 years after the premiere of Lightning, my colleagues still loved working in Classic because they could not get used to Lightning. Was it just about getting used to the new solution? Not only. Certain features are available only in Classic.

Let me describe them to you. The first one is dashboard scheduling: scheduled refreshing of dashboards and sending them by email, just like scheduled reports. Cool, right? With list view sharing in Lightning, you have the possibility to set visibility for a list view for yourself, publicly, or for roles/subordinates, whereas, in Classic, you had the possibility of sharing a list view for a group of users. Another is JavaScript buttons. Salesforce Classic supported JavaScript buttons, which allowed building logic in Java that was triggered by clicking a button. These are the three most important elements that remained in the old version, slowly already forgotten, because who uses Classic nowadays? (If you use it, raise your hand!).

In summary, the appearance of Salesforce Lightning echoed widely in the Salesforce world. It was not the introduction of, for example, Dynamic Forms, which are great, or new features for FLOW. It was a completely new interface with many conveniences. It's as if Marc Benioff at the latest *Dreamforce* presented Lightning+ with a new look and functions; it would take us a moment to get used to it, wouldn't it? But there's no Lightning Experience without Lightning Components. Developers with their skills can do the magic using these components. See the next section of the chapter.

Introducing Lightning Components

As the name suggests, Lightning Components are closely related to the Lightning interface. It's a framework offered by Salesforce, which allows developers to create custom-tailored tools for the system user. Similar to fields on objects or automation, components can be customized to the user's needs. Imagine you are the owner of a company selling music for ads on *YouTube*. This music is sold on a licensing basis, meaning the creator can buy it forever or for the duration of specific streams (for example, music for Halloween or Christmas, although the latter reminds us only of Wham!'s *Last Christmas*. Sorry – now it will be playing in your head). So, the owner needs to create a component that will automatically distribute the number of installments considering the type of license (Lifetime or Period), dates such as start/end date, and the total amount to eventually obtain monthly payments for the license. This type of logic in Salesforce OOTB would be possible to do, but it could take much more time than building a component with such functions. Thanks to Lightning Components, a developer can create logic that will create such components with a given logic. But that is not all – the component can look the way the Product Owner wants it to. Layouts, buttons, or even color schemes can be composed to meet their needs. Most elements found in Salesforce are original works of developers from this platform. Among the components, there are Events. No – I am not talking about those events you can find on page layouts of most objects; we are talking about mechanisms that communicate between components and enable the creation of integrated and interactive user experiences. To visualize this more, we can use an example from a company that uses Service Cloud and collects fault reports on its product. The company is called *CleanYourParty* and helps clean up after parties. When the agent picks up the phone, he wants the client to have everything updated. He finds the client and finds a list of reports. After clicking on a specific button, details (Lightning component) and Ticket History (Lightning component) are to be updated. Thanks to this one action, all components are updated. I will not be scattering this in code for you – as an admin, you will not deal with it – but such links exist and allow us to use a system that can communicate. The last element is Services, which is nothing but an interesting toolbox containing tools for testing or debugging solutions.

What to do to create such a component? First of all, know about programming. So, if you get bored with the admin path, feel free to start a developer career. What are the steps?

1. The first is to define the attributes of the component; such attributes are usually agreed upon with the Product Owner as they know what they would like to achieve in the final stage of the product.

2. Next is creating HTML markup, because it must have a skeleton, like everything in the IT world and beyond. We are getting closer to the end. For business logic in JavaScript, we must create a given process – it is a bit like a Sales Process in Opportunity; we need to know what and how it should work.

3. And the last is just a pleasure – styling in CSS – because the component must look good (well, maybe not *must*, but it would be nice if it did).

What after that? Testing. Usually, this is not done by developers, but by **quality assurance (QA)**, people who will spot every smallest error and then send it for correction with a smile on their faces. (Maybe I'm exaggerating with the smile because it's still one team. But thanks to them, we can be sure that the component works correctly.) And debugging – thanks to the appropriate tools, we will be aware of errors in the code.

Lightning Components are a powerful tool, and thanks to them, Salesforce is the number one CRM. Developers work wonders in these components, and as long as the requirements are presented clearly by the client, that is 100% success.

But when the component is completed, how to place it in front of your user's face? For that, use the Lightning App Builder or Classic page layout, more about which you will learn in the next section.

Customizing Lightning pages – Lightning versus Classic

Over the years, Salesforce has repeatedly modified its products to achieve even better solutions. One of the significant ones, as I already mentioned, was the revolution in the form of the emergence of Lightning Experience. In this part, you will learn about two entirely different faces of modifying page layouts in Salesforce. So, take a deep breath, and three, two, one – we dive into page customizations.

In the beginning, there was Classic; that is how we could start writing this book. Administrators, who functioned back when Lightning was not yet introduced, surely had a lot to learn after the premiere of Lightning Experience. Classic, as I already mentioned in the previous sub-chapter, was like the cornerstone for the entire software.

Although Lightning has been on the market for a while now, some companies remained with the old system interface. You might ask why:

- Firstly, it is the cost of migration. The introduction of the new interface also involved migration between the two systems. Therefore, companies decided to stick with the current one, not having to spend a single cent.

- Secondly, stability and reliability (according to the customers who stayed with Classic). Companies using the system for many years; if they did not have any performance issues during that time, why change something that works well?

- Thirdly, developers. It is precisely they who would have to redo and adapt their current system to Lightning Experience. And they usually are occupied with other important projects.

- Fourthly, user habituation. Each of us has a certain routine in the morning: brushing our teeth, coffee, taking the dog for a walk, and working. And what if someone told you to change this order? It would probably take you a while to get used to it. The same with the Classic interface – companies and administrators knew that it would be hard to convince users to instantly switch to the new interface, so they chose a path of slow change or no change.

With the arrival of Lightning Experience, a very crucial question arose: *How can I modify the page layouts of my objects?* The savior was Lightning App Builder. It is a very intuitive tool that should be used to build a fresh look for the records of a given object. Thanks to Drag & Drop, meaning clicking on a field in the list (do not let go), dragging it to the desired place on the layout, and dropping it, modifying pages became childishly simple. The most significant advantages of the new solution were elements such as the following:

- A visual editor, which allowed for real-time preview. You grab a component or field and drop it in the required place. And most importantly, you do not have to visualize it yourself; the system will do it for you.

- Page templates, this functionality enabled the creation of a coherent system, in which pages do not differ significantly from each other in structure.

- A component collection is a list of components available on the left side of the interface. You can go there, find the component you are interested in, and place it in the appropriate spot. Thanks to the division into standard and custom components, we will surely not make a mistake in the choice.

  - Probably, the first time you saw Salesforce, you thought: *This will be interesting, but how to customize it?* And then, the time came when you went to **Setup** -> **Object Manager**, selected an object, and started making changes. I remember my first changes on the dev org, and I thought it would be easy to do. But as it is with everything, it was just the beginning; the changes I made were the first steps in being an administrator in the system. If you have trouble accessing the Classic page layout, let me help you. Here, you'll see how to do it:

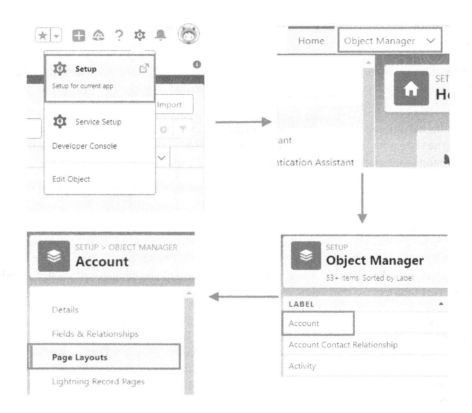

Figure 4.4: Four steps to finding the Classic page layout

- Upon selecting the page layout, you can find all fields that have ever been created. So, if you create a new field and want to place it on the Classic page layout, look at this list. The helpful quick-find window will assist you many times in your administrator life. Among the available elements, you will also find buttons, custom links, quick actions, mobile and lightning actions, expanded lookups, related lists, reports charts, visual force pages, and Custom S-Controls. Yes, I know – it is a lot. After adding your field, and hovering over it with the cursor, you can see two icons – **Remove** and **Properties**. I do not need to explain the first one, while the latter is used to set the field as **Read-Only** or **Required** (meaning, you cannot proceed without a value in this field). A few important visual cues – fields with a blue dot are the ones you cannot remove from the page layout, those with a red asterisk are required, and those with a padlock are read-only. The Classic page layout allows you to divide it into sections, dedicated to business needs.

At the bottom of the page layout, you will find all related lists (that is, objects in relation and their records). If you click on the small tool icon, you will get to related list properties where you can add displayed fields and additional buttons.

And now, *the king is dead, long live the king!* A bit about Lightning App Builder. Just like in Classic, I will show you how to get there:

1. Click on the gear icon in the upper-right corner. Upon opening the drop-down list, select **Setup**.

2. Click on **Object Manager** located in the upper tab.

3. A list of available objects will appear; select any.

4. Now, click on **Lightning Record Pages**, select one of the displayed pages, click on it, and click **Edit**:

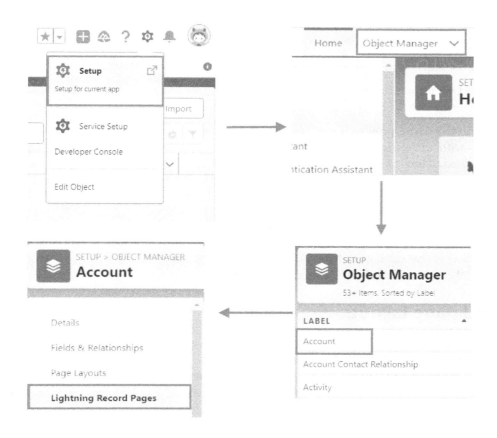

Figure 4.5: Four steps to finding Lightning App Builder

Simple, right?

That is exactly where you can go and make all kinds of changes to the Lightning layout. Upon opening, you will see that on the left side, there is a list of standard and custom components. Each of them contains a brief description that will display when you hover over that component. In the tab next to it, you will find fields that you can place on the layout by dragging. On the main page, what is displayed is what we have on the record page. That is the current view of the Lightning page layout. This is where you can drop new fields or components if you want to add them to displayed ones. When you click on a given component, the properties of this component will display on the right side, where among others, you can set visibility based on specific conditions or other properties.

Lightning pages are responsive, meaning that they adjust to the size of the browser window. So, we know that the page will look good on a mobile phone, tablet, and laptop. We can reproduce such dimension changes using the picklist located at the top of the page with a small monitor icon labeled **Desktop**. Upon selecting the **Phone** option, you will see how the layout changes to a narrow strip representing the record view on the phone. When you finish all the changes, do not forget to save the changes, and activate them. During this second step, you choose the layout availability on three levels:

- **Org Default**, which sets the layout globally for everyone in your entire org
- **App Default** sets the page layout for one selected application – for instance, a layout with service data for the Service app, and with sales data for the Sales app
- **App, record type, profile** – These combined elements will give us, for example, access to service data without details, in the Sales application, for the sales department to see the number of reported claims

It is undeniable that the new will always outsell the old. Therefore, the facilitations in Lightning App Builder will surely evolve for the better. The dynamism that is contained in the new components makes life easier not only for us (admins/consultants), but also for users. But as in life, there will always be enthusiasts of old classic cars and old classic solutions in CRM. Therefore, familiarize yourself with both interfaces so that you know what is better for you. We will start with the Salesforce Classic page layout, and then we will talk about Lightning pages. Let's go!

Record page layouts

In the previous chapter, you learned about the differences between Lightning and Classic pages. You might be inclined to think that this book should exclusively focus on Lightning, as it represents the present and future of Salesforce. However, the current situation is more complex. There is still a significant presence of Salesforce Classic in the Lightning world. In simpler terms, some Salesforce features remain rooted in the old ways, and you will find yourself needing to use them to configure your Salesforce org. In more simple terms, let us say that Salesforce wants us to believe that *Classic never goes out of style*!

One such feature that currently is mixing Classic and Lightning flavors is the Salesforce page layout. Of course, Salesforce is gradually phasing out Classic page layouts, replacing them with Dynamic Forms and Dynamic Actions; however, there are still some limitations that these new Salesforce features are grappling with at the moment. Therefore, you may still need to use Classic page layouts from time to time.

Because, as we just said, today's Salesforce environment is a blend of both Lightning and Classic interface features, we will cover both the Lightning and Classic perspectives to describe how record page layouts are built and configured. Also, for the sake of this book, I will use a theme record page layout or page layout for both Lightning and Classic interfaces. This is just because in both interfaces there are just record pages, and they have their layout, which is built from fields and other data.

What is a Salesforce page layout? So, in both interfaces, Classic and Lightning, the record page layout is a feature that helps you to decide what the Salesforce record user interface will look like. In simple words, the Salesforce record page is just a web page, and editing the Salesforce page layout gives you the power to decide how this page will look. But do not worry – you do not need to know HTML, CSS, or JavaScript to create or edit those Salesforce pages. No matter if you are using a Classic or Lightning user interface, Salesforce serves you with a simple drag-and-drop interface that helps you to decide what, for example, Leads, Contacts, or Account records will look like. So, how to use this feature? Well, the details depend on which Salesforce user interface you are using – Classic or Lightning. But what is important in both interfaces is that you do not have to start from scratch to create a record page interface. Each Salesforce object, such as Leads, Accounts, Contacts, Opportunities, Orders, or any other standard or custom object, will have a ready-made page layout. It is a kind of active basic version that you can use to extend its capabilities or clone it to create a new object's page. So, what can you do with a prebuilt record page layout, no matter if you are using a Lightning or Classic user interface? Let us take a look at a few key features:

- **Adding new fields to the canvas**: In addition to fields that are automatically added to the page layout, such as record name, creation date, or owner, you can also incorporate other fields created within objects.

- **Creating new field sections**: Fields can be organized into sections, enhancing the readability of record pages for users.

- **Integrating new actions or buttons**: You have the flexibility to craft custom actions such as **Compose Emails** or generate object records. These actions or buttons can be positioned on the canvas to serve specific purposes.

- **Incorporating new record page components**: You can introduce new components that have been coded by developers. In the Classic user interface, you can integrate Visualforce components, and in Lightning, you can utilize Lightning components. These components, crafted by your Salesforce developers, can be placed on the record interface to fulfill various business needs.

- **Adding related lists to the page layout**: Objects may have related records that can be displayed here. When an object has lookup or master-detail relationships with other objects, child records can be shown by using related lists on the page layout. For instance, on an Account page layout, you can view Contacts related to that specific Account. We will delve into related lists in the next section of this chapter, so stay tuned for more!

Okay – so we understand what can be done on the record page layout, but knowing how to do it is a different matter. To explain this, we now need to address the Lightning and Classic user interfaces separately. Let's begin with the Lightning user interface.

Working with existing Lightning page layouts

Let's now see how to work with Lightning page layouts:

1. To edit a Lightning page layout, you need to access the Lighting App Builder on the object that you want to set up. Just open any record of this object, click the **Setup** button located on the top of the page, and choose the **Edit Page** option:

Figure 4.6: Opening Lightning App Builder

2. You will see the **Lightning App Builder** interface – Lightning App Builder is a feature created to edit page layouts in the Lighting user interface:

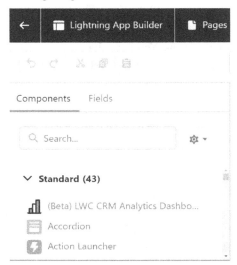

Figure 4.7: Lightning App Builder

3. To add new fields to the canvas, you need to use a Salesforce feature called Dynamic Forms. Dynamic Forms is a feature that transitioned the setup of record field visibility from the Salesforce Classic user interface to Lightning. In simple terms, before Dynamic Forms, you need to edit the Salesforce Classic interface to place or rearrange fields on the record canvas even if you were already using the Lightning interface. With the introduction of Dynamic Forms, you can now do this within the Lightning interface. To utilize this feature, click on the **Record Details** component (the one displaying all the fields in a single component) and then select the **Upgrade Now** action button on the right-hand side:

Figure 4.8: Lightning App Builder – Dynamic Forms

4. Follow the steps on the screen to convert record details to Dynamic Forms:

Figure 4.9: Lightning App Builder – Dynamic Forms conversion

5.   Select which Salesforce Classic page layout will be the source for your Dynamic Forms:

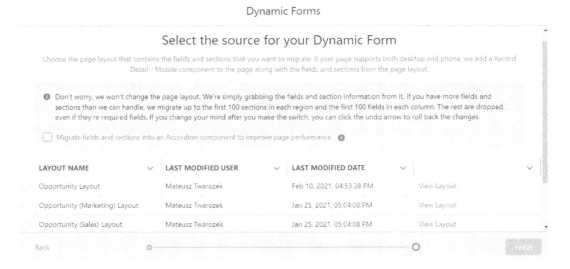

Figure 4.10: Lightning App Builder – Dynamic Forms conversion

6.   After the upgrade, a new subtab called **Field** will appear in Lightning App Builder, and you will see that the component that was displaying all the fields together has changed, and now you will be able to decide about fields separately:

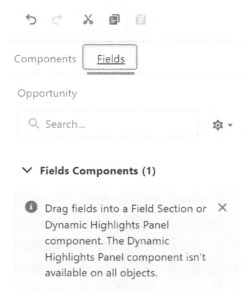

Figure 4.11: Lightning App Builder – Dynamic Forms conversion

7. Here are some things you can do with a field in Dynamic Forms:

- Move on the canvas – place fields in different places on the user interface

- Create field sections and place field/s in the proper field sections

- Set UI behavior – you can set the field to **Read-Only** or **Required**

- Set field visibility – you can decide when the field will be visible for the user. For example, you can display the Account field rating only when the Account type is **Prospect**. Now, this is quite a powerful feature that was not in the Salesforce Classic interface. Before Dynamic Forms appeared in Salesforce, to achieve a similar outcome, you needed to use custom code written by a programmer:

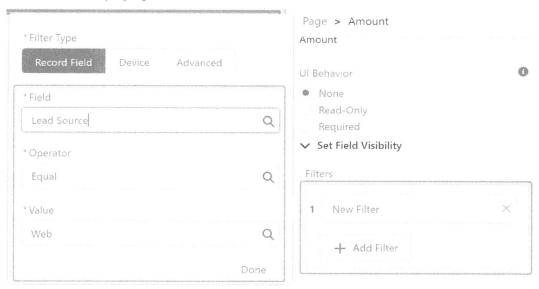

Figure 4.12: Lightning App Builder – Dynamic Forms converted filters

- Use **Tabs** and/or sections to arrange fields, related lists, and custom components on the record page layout – you can set the default tab and order tabs the way you want:

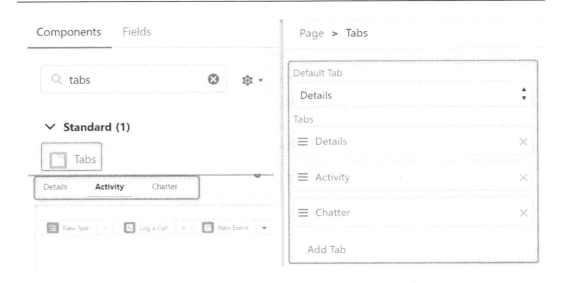

Figure 4.13: Lightning App Builder – Dynamic Forms converted tabs

Sections can be created using the **Field Section** feature, as shown in the next screenshot:

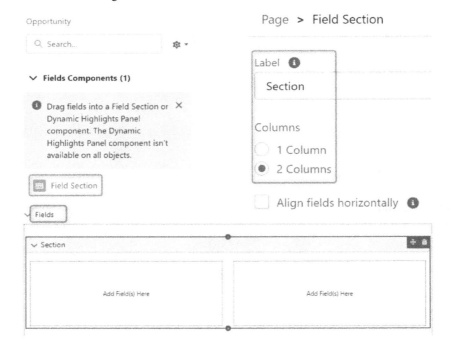

Figure 4.14: Lightning App Builder – Dynamic Forms converted sections

To save your page, just click the **Save** action button on the right-hand side of the user interface. You will be also asked about activation. You can choose the **Org Default** option, and that's it!

> **Tip**
>
> When using **Tabs**, try to "hide" related lists in non-default tabs. This is the recommendation from Salesforce to enhance the speed and performance of Lightning Experience pages. Hiding these lists can speed up page loading because Salesforce won't have to display lists that often contain many records every time you enter the page. Salesforce suggests placing only the most important information in default tabs.

Okay – now that we know how to edit a Lightning record page, what about a Classic page layout? Should you even be concerned? The answer is yes, you should. At least at the time of writing this book, there were still some limitations related to Dynamic Forms on certain Salesforce standard objects. For example, Dynamic Forms aren't supported on objects that are not LWC-enabled. Objects such as Campaigns, Products, and Tasks, which are not LWC enabled, still rely on information from page layouts. Additionally, there is currently no way to configure record header fields using Dynamic Forms in the Lightning interface. To accomplish this, you need to edit the Salesforce Classic compact layout. For these reasons, you will still need to work with Salesforce Classic page layout features from time to time. But don't worry – let's explore together how things are structured there.

How to create (or edit) a new Classic page layout

Just follow these simple steps to create a new Salesforce page layout:

1. In Salesforce **Setup Object Manager**, just access the object you want to create a layout for (for example, Account).
2. Navigate to **Page Layouts** under the object's settings.
3. Create a new layout, specifying the name and layout type (if you want to edit an existing page layout, just click the **Edit** action next to the page layout name).
4. Add and arrange fields in the layout – use the drag-and-drop interface to move fields from the top of the canvas to proper page sections.
5. Add and arrange related lists. In a similar way to adding fields, you can add needed related lists. For example, on the Account page layout, the Contact is a related list that you will use for sure.
6. Assign the layout to specific Salesforce profiles.
7. Save your changes to activate the layout.

As you can see, working with the Salesforce Classic page layout is very easy. Let's now see how to handle Salesforce Classic compact layouts.

Creating a new Classic compact layout

This one is important for the simple reason that the Classic compact layout and the fields defined there are utilized as a header in the Lightning record interface:

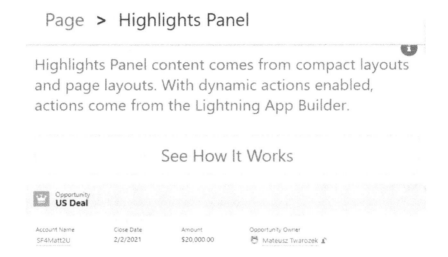

Figure 4.15: Lightning App Builder – Dynamic Forms converted header

Now, let's check how to handle Salesforce compact layouts. Follow those steps to create a new compact layout that can be used as a Lightning record header:

1. In Salesforce **Setup Object Manager**, just access the object you want to create a layout for (for example, Account).

2. Navigate to **Compact Layouts** under the object's settings.

3. Click **New** to create a new compact layout.

4. Select the fields you want to include in the compact layout – those fields will be shown in the Lightning record header.

5. Choose Salesforce profiles for which this compact layout should be available.

6. Save your changes to activate the compact layout.

As you can see, creating a Salesforce compact layout is very straightforward. It essentially involves only a few clicks, but the importance of this feature is currently high, and you will certainly be using it shortly.

Now we know almost everything about Salesforce page layouts, as we can handle both the Lightning and Classic versions of the Salesforce user interface. In the next section, we will learn how to make those page layouts richer by adding other related records to them using Related Lists.

Related lists

Related lists is a Salesforce user interface feature that you will use many times. A related list is simply a section of a record's detail page that displays all items associated with that particular record. From a technical perspective, the entries in a related list consist of records that have a connection through a lookup or master-detail relationship with another record. You will use this feature often because having quick access to records displayed on related lists is often very important for you. Quick access to Contacts or Opportunities on Account records is just one of the most common examples of using related lists. I deliberately used the plural "lists" because this functionality appears in Salesforce in various forms. Let's get to know them all.

Here are some related list types in Salesforce Lightning:

- **Related Lists**
- **Related List – Single**
- **Dynamic Related List – Single**
- **Related List Quick Links**

Let's take a look at the following screenshot to see how related lists are visible in Lightning App Builder:

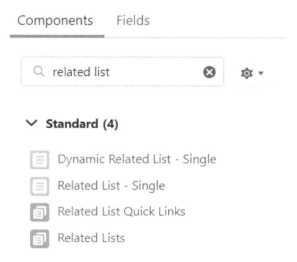

Figure 4.16: Lightning App Builder – Dynamic Forms related lists

Now, let's look at each of them in detail:

- **Related Lists** – The most standard related list component, it's prebuilt in the object's user interface for objects with related records. The **Related Lists** component displays all records related to the object. For instance, on the Account object, it will show Contacts, Opportunities, Cases, and records related to the Account from other objects:

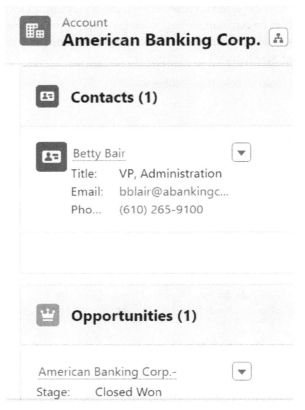

Figure 4.17: Lightning App Builder – Dynamic Forms Related Lists component

You can decide how a list will be displayed by changing the related list type. You can choose between **Default**, **Basic List**, **Enhanced List**, and **Tile**. Just play around with those options to see the differences:

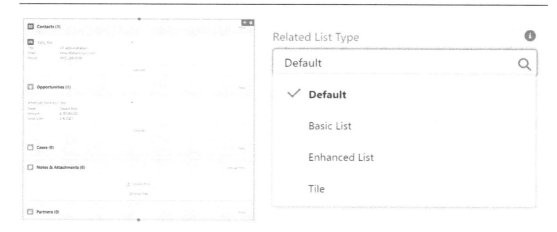

Figure 4.18: Lightning App Builder – Dynamic Forms related list types

- **Related List – Single** – Very similar to **Related Lists**, but it allows you to display records of only one selected object:

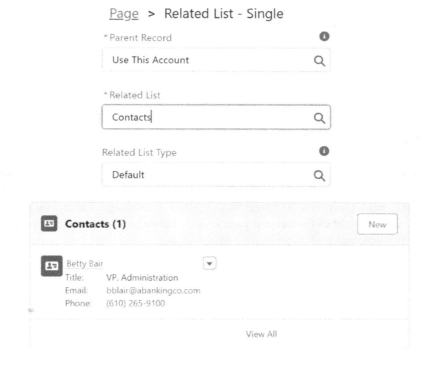

Figure 4.19: Lightning App Builder – Dynamic Forms Related List – Single component

So, the difference between the **Related List – Single** and **Related Lists** components is that the first can show records of only one object while the second can show records of all related objects:

Figure 4.20: Lightning App Builder – Dynamic Forms related lists' differences

I've mentioned several times that the current Salesforce user interface is a blend of Lightning and Classic. Here's another advantage of that statement. Occasionally, when you try to add a **Related List – Single** component, the system will greet you with an alert message: **This related list cannot be displayed because it is not in the current page layout.** Take a look at the following screenshot, which illustrates this behavior:

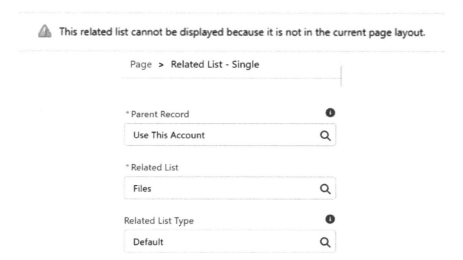

⚠ This related list cannot be displayed because it is not in the current page layout.

Page > Related List - Single

* Parent Record   ⓘ

Use This Account   🔍

* Related List

Files   🔍

Related List Type   ⓘ

Default   🔍

Figure 4.21: Lightning App Builder – Dynamic Forms related lists issue

What does this mean? Have I done something wrong? Why this error? Am I a bad Salesforce admin? Keep your head up – nothing like that! It just means you have to depart from your sweet and funky Lightning user interface and switch to the old-school Classic page layout editor. You'll need to edit it to include the required list view. A bit weird, I know, but hey, that's just how Salesforce rolls at the moment! It's like leaving the party in a fancy suit and coming back with a classic cape – a bit wacky, but we're rolling with it!

- **Dynamic Related List – Single** – Now, that's a relatively new Salesforce feature. As you may have noticed, new Salesforce features related to the user interface often have names that include *Dynamic*. **Dynamic Related List – Single**, as the name stands, is capable of displaying records related to a single object similar to the **Related List – Single** component. What more do you need than that? Let's describe new features related to this version of a related list:

  - **Related List Label** – You can rename the list. For example, you can rename **Contact** as **People** or **Employees**.

  - **Related List Fields** – You can add existing fields to the list, change the order of columns, and hide fields.

  - **Sort Fields** and **Sort Order** – You can choose which field will be used to sort the records in the list.

  - **Related List Filters** – You can filter the records you want to see in the view. Now, this is a very interesting and awaited feature that you will use for sure.

  - **Actions** – You can add action buttons to related lists; for example, the **New Record** button, the **Add to Campaign** button, or any other custom action button.

In the next screenshot, you can see what the **Dynamic Related List – Single** component looks like in Salesforce Lightning App Builder:

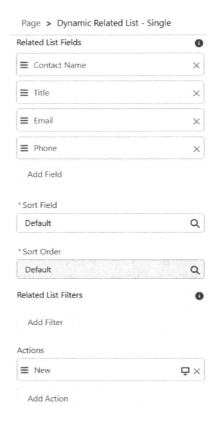

Figure 4.22: Lightning App Builder – Dynamic Forms related lists features

> **Tip**
> In the Salesforce org that I set up, I often use Dynamic Related List Filters to display different lists of won Opportunities and others to show lost and in-progress Opportunities. This way, users have a clear view of both past and current deals associated with any Account.

Okay – so, you are now a master of Salesforce user interface management! Congratulations! Managing the Salesforce UI is an essential skill that empowers you to navigate the platform seamlessly and maximize your efficiency. This proficiency opens the door to unlocking the full potential of Salesforce, enabling you to optimize the way the user uses the platform, track essential data, and be informed about important data that can drive business success.

Salesforce for mobile

In modern times, fast access to data is a primary key to success. Of course, Salesforce recognizes this, which is why there is the possibility to access your organization on mobile devices. Whether you are using an Android or Apple phone, Salesforce has you covered. You can install the dedicated app directly from the Play Store or Apple Store. With the Salesforce mobile app, essential information is always at your fingertips. But what exactly can you do with this app?

Let's list the most important mobile app features:

- Real-time collaboration for efficient teamwork

- Instant access to sales data, records, and reports for quick decision-making

- Accelerated deal closures with progress visibility

- Streamlined daily organization for schedules and tasks

- Priority notifications for urgent messages and approvals

- Easy task management

- Effortless case organization and escalation

We have just listed the main features, but how to set up this app? It's not very complicated, but to make it a little bit crazy, Salesforce has spread these settings to different places in the system. What's more, Salesforce still maintains old and newer mobile solutions for some features – for example, related to navigation items. Okay – but so as not to scare you too much, let's just check what's in the system:

1. You can start by accessing **Setup** where you can configure a few things such as the following:

    - **Branding** – Deciding about your app branding details such as brand color and loading page logo.

    - **Navigation** – Deciding about app menu items. This setup can be done here because users can keep using the Mobile Only app (currently, it can be still used, but it is a legacy app), or they can switch to a Lightning app using the App Launcher. It can also be done in Lightning App Builder.

    - **Notifications** – You can enable in-app or push-app notifications here.

    - **Offline settings** – You securely cache data from Salesforce for Android and iOS on mobile devices, which will result in faster viewing of previously accessed records:

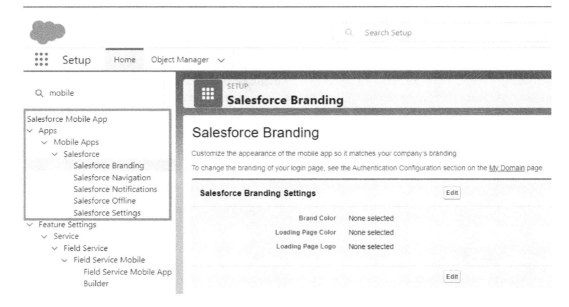

Figure 4.23: Salesforce mobile setup

2. Now, let's see how to give users access to the mobile app. To set this up, you need to go to **Manage Connected Apps**:

Figure 4.24: Salesforce mobile app access

3. Click **Edit Policies** to decide about app access for permitted users:

Figure 4.25: Salesforce mobile app policies

4. If you choose the **Admin approved users who are pre-authorized** option, only users with the associated profile or permission set can access the app without first authorizing it:

Figure 4.26: Salesforce mobile app permitted users

Once you choose this option, you can either handle app profiles by editing each profile's connected app access list or manage app permission sets by editing each permission set's assigned connected app list. (More information available at https://help.salesforce.com/s/articleView?id=sf.branded_apps_allow_deny_con_app.htm

5. Finally, you need to understand how the record page layout works in Salesforce mobile apps. The mobile app record page layout is based on the Desktop page layout, but you know Salesforce – there are a few little crazy key details you need to wrap your head around to get the full picture. Finally, you need also to understand that Salesforce mobile apps are not the same as Salesforce access via phone browser (for example, the Chrome browser installed on your phone). The mobile app page layout and mobile phone form factor are set up differently. Mobile apps sometimes have access to different features than web versions or show the data in different ways. The list of those differences is quite big, and it may change from time to time with each new Salesforce release. (For more details about what's different or not available in the Salesforce mobile app, visit the following web page:

```
https://help.salesforce.com/s/articleView?id=sf.limits_mobile_
sf1_parent.htm&type=5)
```

If you would know more about Salesforce mobile app customization, you can visit the Salesforce Trailhead platform and look for the dedicated modules.

Summary

Congratulations to you, our valued reader, for making it this far in your Salesforce adventure! Thanks for sticking around and keeping the lightning in your learning journey! Let's summarize now what you have learned in this chapter.

In *Chapter 4*, you delved into the dynamic realm of Salesforce user interfaces. You started by exploring the transition from Salesforce Classic to the Lightning interface, uncovering the major changes and results of this shift. You also ventured into the world of Lightning components, discovering how these building blocks could enhance your Salesforce experience.

The chapter also took a closer look at customizing Lightning pages, highlighting the differences between Lightning and Classic interfaces. You explored page layouts and the role they played in optimizing the user experience.

In the section on related lists, you examined how to manage and display related records efficiently. Finally, you introduced Salesforce for mobile, shedding light on the unique considerations and opportunities for mobile device users. With these insights, you'll be well equipped to navigate and customize the various Salesforce user interfaces, ensuring you harness the full power of Salesforce. Knowing so much about user interfaces, you can now confidently enter the realm of Salesforce objects in the next chapter.

Further reading

To read more about Salesforce mobile features, please visit the following web page: `https://
developer.salesforce.com/docs/atlas.en-us.218.0.salesforce1appadmin.
meta/salesforce1appadmin/sf1_features_to_app_comparison.htm`

5

Objects in Salesforce

In this chapter, we will focus on what is exceptionally important in the Salesforce structure, namely objects. As I mentioned earlier, Salesforce may remind you of Microsoft Excel, and I believe this software can be dubbed the precursor to most CRMs worldwide. So, if we draw a comparison to Excel, the object discussed in this chapter is akin to a spreadsheet. It's the place where we keep our columns and rows of data. This is a structure that enables data storage. That's what an object is, and that's the end of it. Of course, I am joking – this is just the beginning. In this chapter, you will learn the differences between standard and custom objects, as well as what junction and external objects are.

I invite you to the world of objects, elements without which Salesforce wouldn't exist. Here are the topics you will find here:

- Standard/custom objects
- Junction objects
- External objects

Standard/custom objects

In the previous chapters, you learned about the components of Salesforce. Indeed, it has meticulously written lines of code that look awesome after graphic processing. You also learned how the system is divided among its tenants, and what Classic and Lightning are. As you can see, some of the Salesforce elements are duplicated. The same goes for objects – we have standard and custom ones. Objects in this ecosystem are an absolute must-have, and it's these two types you will learn about in this section. So, let's get started.

To better illustrate the meaning of an object, imagine two warehouses with thousands of different products. In the first one, James, the chief warehouseman, is a disorganized person. Papers are scattered under his feet, products are strewn around the warehouse, and when an order comes in, he scratches his head, trying to find the right goods. On the other hand, in the second warehouse, there's John. John loves order; he knows which box contains what. When an order comes in, it takes him 5 minutes to complete it. But getting back to objects. They are like John's warehouse. Thanks to their fields, and properties, they can hold the relevant data. Entering the **Contact** object, you know well that you will find data such as first name, last name, or email address. And entering the product, you know that there will be product data such as the name and other elements associated with it.

If we already know what objects are, we can move on to their classification. In the Salesforce system, we have access to four types of objects. However, in this section, I will focus on two, which I believe are the most popular (you will find the rest in the next sections) – standard objects are those that we receive in the out-of-the-box version, which is available in a fresh Salesforce org. These objects have been designed to best reflect common business constructs. They are designed to represent sales processes and service processes. As a result, people who have been working in the Salesforce ecosystem for a longer time can offer their clients Salesforce clouds tailored to their needs. Standard objects have implemented the most necessary fields, which makes the implementation of CRM solutions relatively quick.

The most popular objects include the following:

- **Account**: Here, users store data of companies or organizations. Here, you will find the company's location, email address, or publicly available phone number.

- **Contact**: This object stores data of people, often employees of a given company. In this object, you will find the first and last name, phone number, and date of birth.

- **Opportunities**: In other words, chance for a good deal. There users store data about the sale of a service/product. The name of the opportunity, the type of customer, the amount, or the close date are the fields that will be on the OOTB version of an opportunity in Salesforce.

Of course, I am talking here about basic Salesforce with Sales Cloud. If you switch to Service Cloud, you will find other objects dedicated to service processes.

Each of the standard objects can be customized to the needs of your company. For this, we use new fields that we can place on the object. We talked about fields in *Chapter 3*, remember? Thanks to this type of customization, we can even adapt standard objects to our needs.

As I mentioned, objects are a kind of container storing appropriate data. This data can be combined using two quite interesting, quite similar, but also quite different connections.

The first one is the **master-detail** relationship. This relationship is between two objects, and what is characteristic is that in this type of relationship, we can establish a hierarchy and dependency. Here, we set up a parent-child relationship, where the master object is the parent and the detail object is the child. This type of relationship has certain dependencies; the records of the detail object are always controlled by the records of the master object. To better understand, look at the following screenshot. We will use the entire community that loves events such as *Comic Con*. As you can see, in the next diagram, they are waiting for their special event, but at the last minute, the organizer cancels the event. And all the regular attendees go home:

Figure 5.1 – Children (attendees) with parent (special event); no
children (attendees) without parent (special event)

Based on this example, if the parent in a master-detail relationship is deleted, the child records (detail-child) will also be removed. This is the opposite behavior to that in a lookup relationship. But about that in the following sections.

But are these all the important aspects of this relationship? Of course not. Another extremely important element is security. In a master-detail or parent-child relationship, the child inherits all security settings from the parent object (it's a bit like a family; you can't fool DNA). This means that the security and visibility settings of the master record object enforce the same rules on the detail records. And the relationship on the detail object is necessary to create a record. Without specifying a master record, you cannot create a new detail.

> Tip
> If you want the ability to use roll-up summary fields, remember that they only work in a master-detail relationship. These fields are not available if you want to use them in a lookup relationship. Why is this the case? Because they can aggregate data from the child detail records and present summarized values on the master record. For instance, this could be the total of orders (detail) for a customer (master).

And now, I'll show you how to create such a relationship.

Before proceeding with any configuration, you need to design your structure. Determine which of the objects is master and which is detail. If the objects do not exist, you must first create them:

1. We go to **Setup | Object Manager**, then select the object on which we want to introduce such a configuration. And let's pause here for a moment. Remember to create the master-detail field on the detail object. It is from it that we will create the relationship to the master.

2. We enter the **Fields & Relationships** section, then click on the **New button** and select **Master-Detail Relationship**.

3. Now, we have to select the master object. We choose it from the available **Related To** list. After choosing, we click on it and confirm with the **Next** button:

New Relationship

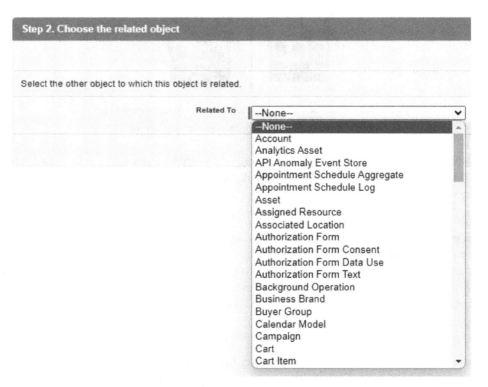

Figure 5.2 – "Related To" dropdown

I. We enter details such as **Field Label** and **Field Name** for the new master-detail relationship. As with any new field, we can add help text and also provide a description. Here, we can also assign a new name for the **Child Relationship** field – an internal name, which if changed can generate errors if, for example, the field is used in integration.

II.   The next item in the same step is setting **Sharing Settings**, which allows granting access to the master object at the **Read** or **Read/Write** level. And we set **Allow reparenting**, which is nothing but changing the master object for the child.

III. The last setting in this step is the lookup filter, where we can set certain conditions that will apply when selecting a record (for example, if we are looking for an Account, we can set a condition so that records with a billing address in the United Kingdom can be selected). Next, you will see a form for *step 3*, where you will find all these settings:

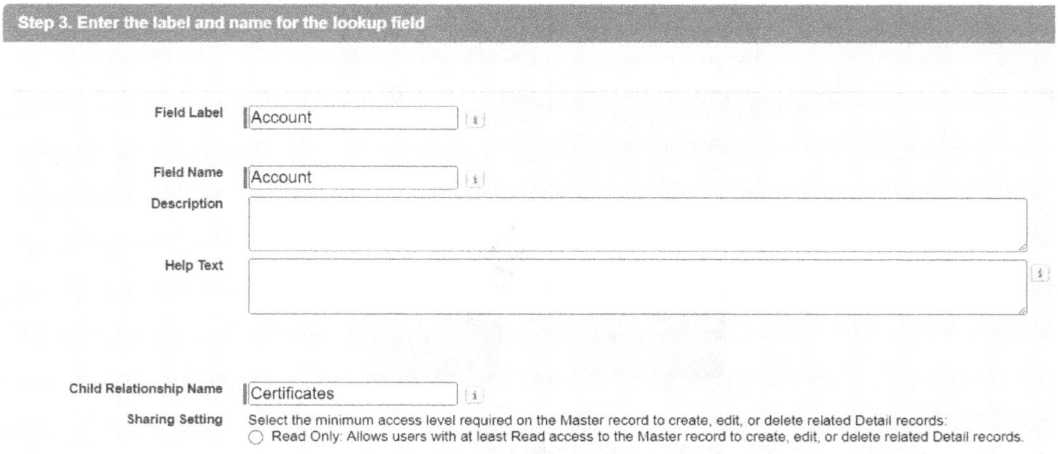

Step 3. Enter the label and name for the lookup field

Field Label	Account
Field Name	Account
Description	
Help Text	
Child Relationship Name	Certificates
Sharing Setting	Select the minimum access level required on the Master record to create, edit, or delete related Detail records:
	◯ Read Only: Allows users with at least Read access to the Master record to create, edit, or delete related Detail records.

Figure 5.3 – Step 3: settings and conditions

4.   The fourth step is fairly standard when adding fields – setting visibility and profiles.

Next, we add the fields to selected page layouts; you will have to choose those on which the master and detail fields are to appear.

And almost finished, because we need to save this configuration with the **Save** button.

Congratulations – you have just created your own master-detail relationship! Unless you just received a **Cannot Create Master-Detail Relationship** error. The reason for this is existing records. Remember when I mentioned that the master-detail relationship is mandatory? That's exactly why we cannot save this field. But fortunately, there is a solution: just create a new lookup field, then add a relationship with the selected object (future master object), and populate the master record in the lookup field. After entering this data into each record, we change the type of field from lookup to master-detail. Voilà – it's ready!

And speaking of the lookup field, it's the more relaxed brother of master-detail. In this relationship, we can connect two records. Thanks to this, users can choose a record from another or even the same object, thus creating a relationship between the two. The records are somewhat dependent on each other, but most importantly, they are not dependent on each other in terms of existence or access. To illustrate this better, look at the next screenshot. If we compare this with the master-detail relationship, in this case, if the *Comicon* event is canceled, but there are plenty of other cool events nearby, the attendees won't go home but will choose another one:

Figure 5.4 – Lookup with record; lookup without or with another record

As you can see, the lookup relationship can manage without the chosen record (unless you set it as mandatory). If the record from the lookup is deleted, nothing happens because the main record can still exist normally in the system.

How to create a lookup? It's quite simple:

1. The principle is quite similar in both relationships (lookup/master-detail). We go to **Setup | Object Manager** and select the object where we want to create a new lookup relationship.

2. We enter the **Fields & Relationships** section, click on the **New** button, and select **Lookup Relationship**.

3. Now, we need to select an object. We choose it from the available **Related To** list. After choosing, we click on it and confirm with the **Next** button.

4. We assign a field label, field name, description, and help text if needed. We enter a child relationship name (internal name). We set whether the field is to be mandatory or not, the behavior if the lookup record is deleted, and add the field to **Reports**.

5. At the very bottom, we have the choice of lookup filter, which is setting the appropriate conditions that will decide which record we can select and which we cannot. It looks like this:

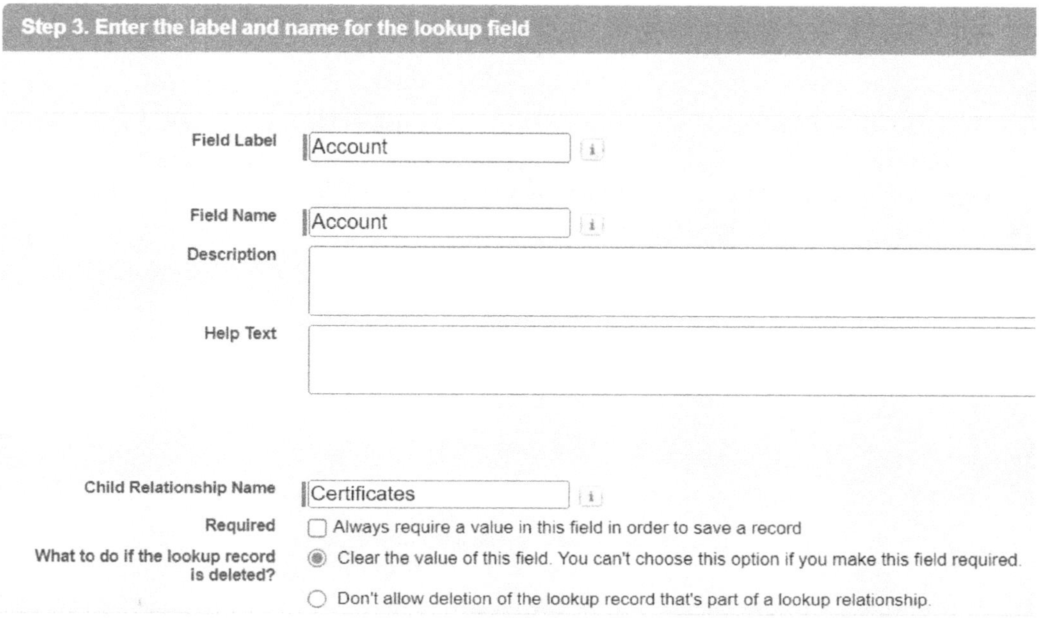

Step 3. Enter the label and name for the lookup field

Field Label | Account | i |

Field Name | Account | i |

Description

Help Text

Child Relationship Name | Certificates | i |

Required   ☐ Always require a value in this field in order to save a record

What to do if the lookup record is deleted?   ◉ Clear the value of this field. You can't choose this option if you make this field required.

○ Don't allow deletion of the lookup record that's part of a lookup relationship.

Figure 5.5 – Lookup configuration settings

6. The fourth step is fairly standard when adding fields – setting visibility and profiles.

7. Next, we add the fields to selected page layouts and save the configuration with the **Save** button.

But what happens when our company is so unique that it needs something tailor-made? Remember those toys where you had to match the right shape with the right hole? When the shapes matched the holes, you knew everything would go smoothly, but what if someone handed you a block of an undefined shape that couldn't fit into any hole? Then, new possibilities needed to be created. This is what custom objects are for. Salesforce couldn't predict every scenario (unless it was like Dr. Strange in *Avengers: Infinity War*), so it gave administrators the ability to create their own solutions.

Administrators can create any object that will store detailed and unforeseen data by Salesforce. These objects are fully customizable to the needs of the company, allowing them to be freely named, equipped with required fields, and to create relationships with other standard or custom objects. Importantly, these objects are scalable, meaning they can grow with the needs of the company. Often companies start with very simple processes that cover a few steps and a certain amount of data. But as companies grow and their needs change, processes also change. Thanks to custom objects, Salesforce also changes and evolves with the client. Such objects grow both in terms of data storage, meaning a larger number of fields, and the complexity of relationships in the system (more master-detail or lookup relationships).

Now that we know what a custom object is, let's check how to create one:

1. Can you guess the first step? Yes – it's **Setup | Object Manager**.

2. Then, click the **Create** button. You will get two options: **Custom Object** or **Custom Object from Spreadsheet**. The latter option allows for quicker creation of a new object using a spreadsheet. But we will go through the first option, which is manual creation. It looks exactly like in *Figure 5.6*:

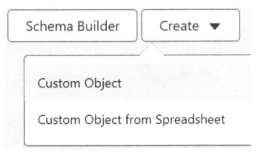

Figure 5.6 – Creating a custom object

3. After selecting **Custom Object**, you will be taken to a form where you need to fill in details such as label, plural label, object name, and description:

Custom Object Information

The singular and plural labels are used in tabs, page layouts, and reports.

Label		Example: Account
Plural Label		Example: Accounts
Starts with vowel sound	☐	

The Object Name is used when referencing the object via the API.

Object Name		Example: Account

Description

Context-Sensitive Help Setting   ◉ Open the standard Salesforce.com Help & Training window
⃝ Open a window using a Visualforce page

Content Name   --None-- ⌄

Figure 5.7 – Custom Object Information

4. Below the **Custom Object Information** and **Enter Record Name Label and Format** sections, you will find additional options. Thanks to checkboxes in the **Optional Features** section, you can immediately add the object to reports, enable activities related to it, or enable history tracking.

5. In **Object Classification**, you can enable or disable sharing of the object, as well as Bulk/Streaming API options.

6. Next, you have the option to set the deployment status. **In Development** will not share the newly saved object, keeping it on the backend without user access. The **Deployed** option will provide access to the new object.

7. The last two sections are equally important. The first is **Search Status**. This option allows you to enable internal Salesforce search for the new object. The last option gives you the ability to add notes and attachments to the new object on the page layout. And one of the more important checkboxes is the last one. It activates the **New Custom Tab Wizard**, which is a form that allows you to create a new tab. Without this, you will have to do it yourself through an option located elsewhere.

8. After selecting all the options you want in the new custom object, click the **Save** button. This will save your new object, in which you can start creating new relationships, fields, page layouts, and much more.

> Tip
>
> Important question: Do you know how to recognize that an object is not a standard object, but a custom one?
>
> a) It has the name **New Awesome Custom Object**
>
> b) It blinks green
>
> c) Right next to the name, a little GIF of a dancing unicorn appears
>
> d) It has an appendage in the API name: __c
>
> Of course, the answer is d). Although I very much regret that there is no dancing unicorn. I don't know about you, but I can already see it in my head. There's another way – you simply look at the type of object in **Object Manager**, and there it's written whether it is a standard or custom object. But that sounds too easy for us, right?

The last thing I would like to share with you in this section is best practices when creating a new object:

- The first is naming. Remember to make the name reflect the function of the object. It is very important that the user/administrator/developer knows what the new object is for. Avoid abbreviations.

- The second very important element is descriptions. Descriptions can save many a new administrator. Thanks to descriptions, a new employee can easily find their way in a new structure for them.

- Third is configuration. Design your object in your head, and think about sharing the object and access to its data. Your fields that will be mandatory and validations for the rest of the data will prevent chaos among the data, and you will take care of its quality.

- Fourth is documentation. No matter what you use, whether it is Confluence or maybe a document shared on Google Drive, fill it with new fields and settings. New people will be very grateful to you.

And so, we have come to the end of the section on standard and custom objects. After this chapter, you will know what they are and what they can be used for, how to create connections between such objects in two ways (master-detail and lookup), and how to recognize a custom object. In the next section, you will learn about the mysteriously sounding junction object.

Junction objects

In the previous section of this chapter, we learned what a Salesforce object is. We also observed that objects can have relationships with each other, such as lookup or master-detail. Both relationships are crucial, as most of the time, you'll use them to connect Salesforce objects. Thanks to these relationships, you can effortlessly establish one-to-many relationships where there's a single parent, but multiple child records can be associated with it. However, there will be instances where these types of relationships might not be sufficient to meet the business requirements you'll encounter.

If you've spent some time working with Salesforce Accounts and Contacts, you may have noticed that it works seamlessly when connecting a Contact to a single Account. On the Contact record, you'll observe that a specific Contact is linked to one and only one Account. But what about scenarios where this Contact is associated with multiple Accounts? This can happen when an individual works for various companies, known as Salesforce Accounts.

To establish this type of connection, we need to create not just one-to-many relationships but many-to-many relationships. In simpler terms, one Contact can be linked to multiple Accounts, and conversely, one Account can be connected to several Contacts. Salesforce has provided a solution to this challenge by introducing an additional object called **Account Contact Relationship**, which establishes connections between Accounts and Contacts, opening the door to creating many-to-many relationships. Objects of this kind, connecting two other Salesforce objects and forming many-to-many relationships, are referred to as junction objects.

Let's consider an additional example. Suppose you want to create a recruitment app within your Salesforce org. The app should store information about recruitment projects and candidates. Of course, you can see that to build this solution, you would need to use two objects: Recruitment Project and Candidate. You also need to connect them somehow to assign a candidate to a specific recruitment project.

Now, let's explore if we can use standard relationships such as lookup or master-detail to connect these objects effectively. As you may recall, lookup and master-detail relationships offer easy possibilities to create one-to-many relations. This means that using these types of relationships, we could connect a recruitment project (one) to candidate/s (many), like Salesforce connects Accounts and Contacts. So, in one recruitment project, we could have many candidates assigned. Everything seems nice and straightforward, right? Unfortunately, no. In the real world, one candidate can be a part of not just one but many recruitment projects. For this reason, the one-to-many relationship is not sufficient.

To build your solution, you need to create a junction object that will connect candidates to many recruitment projects, like how Salesforce serves you when connecting one Contact to many Accounts. Let's call this junction object Candidate Application. Now, let's see which objects we finally have: Recruitment Project, Candidate Application, and Candidate, where Candidate Application acts as the object connecting Recruitment Project and Candidate.

Now, every time a candidate applies to a recruitment project, they are not directly assigned to it (as this would create a one-to-many relationship), but the candidate is primarily assigned to the **Candidate Application** record, which is then connected to the **Recruitment Object** record. This creates the many-to-many relationships built on top of the junction object.

Okay – but let's stop with theory for now. Let's now see how to build the junction object logic in Salesforce.

How to create a junction object

Let's now use a recruitment app example to see how junction object relationships can be established in Salesforce. Simply follow the next steps to construct the logic for your recruitment app:

1. Create a recruitment project custom object:

   I.   Use **Data Type Text Name** for an object.

   II. Check the **Launch New Custom Tab Wizard after saving this custom object** checkbox:

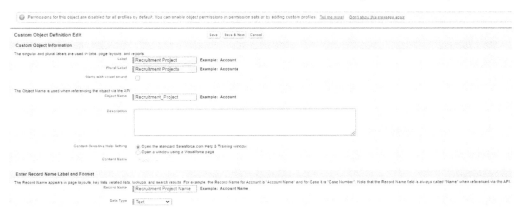

Figure 5.8 – Recruitment Project object

2. Create a candidate custom object:

    I.   Use the auto-number data type for the record name format; for example, CAN-{000000}.

    II. Create two custom text area fields (255 characters each): **Name** and **Surname**:

Figure 5.9 – Candidate object

    III. Check **Launch New Custom Tab Wizard** after saving this custom object.

3. Create a candidate application custom junction object:

    I.   Use the auto-number data type for the record name format; for example. CAP-{000000}.

    II. Create two custom master-detail relationship fields: **Recruitment Project** (related to the Recruitment Project object) and **Candidate** (related to the Candidate object):

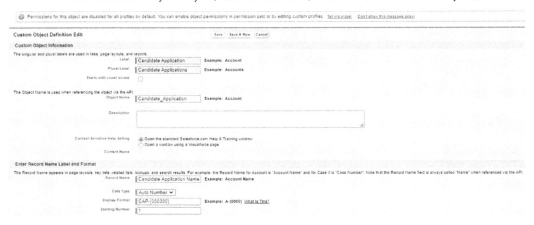

Figure 5.10 – Candidate Application object

    III. Check **Launch New Custom Tab Wizard** after saving this custom object.

4. Update the **Recruitment Project** related list to show the name and surname of the candidate:

    I. Go to the Recruitment Project object in **Setup**.

    Go to **Page Layout**, click on the name of the layout to enter edit mode, and add two fields (columns), **Name** and **Surname**, to the **Candidate Applications** related list to be able to see them in the **Recruitment Project** record. Remember to save your changes:

Figure 5.11 – Recruitment Project object related list update

5. Test your solution:

    I. Create two **Recruitment Project** records:

    II. Go to the **Recruitment Project** tab, then do the following:

    i. Create two **Recruitment Project** records:

        • First project name: **Sales Director**

        • Second project name: **Sales Representative**:

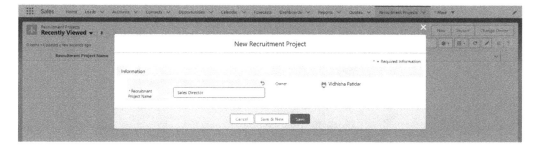

Figure 5.12 – Recruitment Project record creation

    III. Create a **Candidate** record:

Go to the **Candidate** tab, then do the following:

i.   Click **New** to create a **Candidate** record:

- Name: **John Doe**:

Figure 5.13 – Candidate record creation

IV.   Create **Candidate Application** records for each recruitment project:

i.   Go to candidate record **John Doe** (CAN-000001))

ii.   Create two **Candidate Application** records:

- Go to the candidate record **Related** tab.
- Click the **New** button on the **Candidate Application Related** list.
- The **Candidate** field will be already prepopulated.
- Fill in the **Recruitment Project** field. For the first record, choose **Sales Director** and for the second, choose **Sales Representative**:

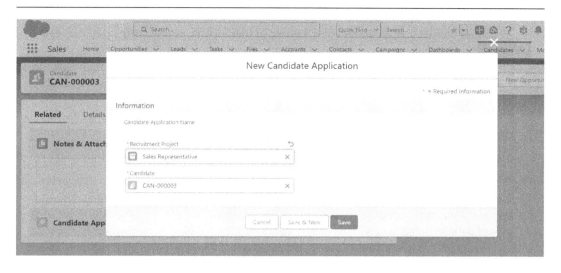

Figure 5.14 – Candidate application record creation

V.   See the results – now, you can see that on two existing **Recruitment Project** records, **Sales Director** and **Sales Representative**, you have two candidate applications (one for each) received from the one and the same candidate, which is John Doe. Simple and funky, isn't it?

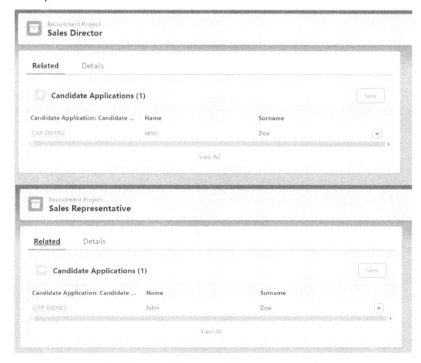

Figure 5.15 – Candidate application visibility in two recruitment projects

You've just acquired the skill to create Salesforce many-to-many relationships using a junction object. This knowledge is incredibly powerful, as the creation of many-to-many relationships is a crucial aspect of numerous business requirements. Now, you are well prepared and fully capable of mastering this task with confidence. In the next section, you will learn about a different interesting type of Salesforce object called external objects.

External objects

External objects function much like custom objects, with the distinction that they are linked to data residing beyond your Salesforce organization. Every external object is dependent on an external data source definition, facilitating the connection to data stored in an external system. Each definition of an external object corresponds to a data table within this external system, while the fields of an external object correspond to the columns within that table. External objects are searchable, and users can interact with this data also. Simply put, external objects are special types of Salesforce custom objects that can display data stored outside of Salesforce. Sounds cool? That's because it is! But what makes this "coolness" even cooler? There are lots of benefits to using external objects. Let's list here the most important:

- **Data storage savings** – External objects prove most advantageous in scenarios where there's a huge volume of data that you either can't or prefer not to store within your Salesforce organization. Sometimes you just do not need to keep millions of records in Salesforce when they are already kept in your other system. You just need to give users access to them directly from your CRM.

- **Quick access to important data** – In most cases, you only require access to a limited dataset at any given point in time. Through on-demand access to record data, external objects consistently mirror the real-time status of external data. This obviates the need for maintaining a redundant copy of this data within Salesforce, preventing the squandering of storage and resources on data synchronization.

- **Better user experience** – The next important fact is that when using external objects, users do not need to go out of Salesforce to check external data. This means fewer clicks for the users and a less distracted user experience.

- **You can create Salesforce reports based on an external object's data** – Yes, you can use Salesforce reporting capabilities to create reports and dashboards that will show your external data. Cool, isn't it?!

Of course, there are some important limitations related to external objects as well. Let's list a few of the most important ones:

- **Security** – You can't use sharing rules to control object access

- **Limitation to 200 objects** – Each organization can have up to 200 external objects

- **SOSL and OData limits** – You need to check existing limits if you are planning new custom development involving external objects

Besides all the advantages and limitations of external objects, there is one general thing that is very important: you will need to pay more to get external objects! Yes, external objects are a Salesforce feature for which you need to pay additionally, and they are available with **Salesforce Connect and Files Connect** licenses.

Okay – so, we already know what external objects are used for and how fun they are. Now, let's see how to create one.

How to create an external object

Just follow these simple steps to create a new external object (but remember that to utilize all its data sync features, you'll need to purchase additional Salesforce licenses):

1. Navigate to **Setup** and use the **Quick Find** box to search for `external objects`. Select **External Objects** from the search results – you can either create a new external object by clicking **New External Object** or edit an existing one by selecting **Edit**.

2. Enter the following details:

    *   **Label** – Any user-friendly name for the external object, displayed in the Salesforce user interface, such as in list views.

    *   **Plural Label** – It will be used as a tab name.

    *   **Starts with vowel sound** – Choose "an" instead of "a" for label prefixes when appropriate for your org's default language.

    *   **Object Name** – A unique identifier that must be unique and follow specific naming rules.

    *   **Description** – An optional description.

    *   **Context-Sensitive Help Setting** – Defines what users see when they click **Help for this Page**.

    *   **External Data Source** – Choose the external data source definition with connection details.

    *   **Table Name** – The external system's table to which the external object maps.

    *   **Display URL Reference Field** – Only for Salesforce Connect, this field is autogenerated from the external system.

    *   **Allow Reports** – Available only for Salesforce Connect.

    *   **Deployment Status** – Indicates visibility to other users.

    *   Launch **New Custom Tab Wizard** after saving – If selected, the custom tab wizard starts after saving the external object and you will be able to create a dedicated tab that can be used by the users to navigate to the external object's data.

    *   **Allow Search** – If selected, you will let users find external object records via SOSL and Salesforce global searches.

3. Save the configuration.

4. On the external object detail page, view and modify custom fields, relationships, page layouts, field sets, search layouts, buttons, and links. To add fields or create mappings, click **New** in the **Custom Fields & Relationships** related list. For different page layouts by user profile, select **Page Layout Assignments**.

You have just learned what an external object is, when it can be used, and how to create one. It's truly valuable to explore the possibilities of external objects, especially as Salesforce, as a CRM, frequently integrates with various data sources (such as ERP systems, billing systems, and so on). Utilizing external objects can be one viable option for building such integrations.

Summary

Thank you for being with us on this exciting Salesforce adventure! You have just finished the next chapter of this book, and your Salesforce knowledge is blossoming. Let's now summarize here what you have learned in this chapter.

In this chapter, you have primarily learned about Salesforce standard and custom objects, gaining practical knowledge on how to create a brand-new custom object. Moreover, you have learned how to establish relationships between Salesforce objects using lookup or master-detail fields.

You've also learned about Salesforce junction objects, their logic, and usage, and even had the opportunity to create one in a real-life example while building a simple Salesforce recruitment app.

In the final section of this chapter, you've gained insights into external objects, their practical applications, and practical knowledge on how to create such an object.

As this chapter concludes, the Salesforce saga continues. Join us in the next chapter where we will discuss Salesforce user management and security.

User Management and Security

Welcome to *Chapter 6*. You are getting closer to the halfway point of the book. In the previous chapter, you learned about the differences between standard, custom, junction, and external objects, and their application in the system. I'm sure this has helped you understand the most important aspects of Salesforce's structure. Now that we have the structure, let's take a closer look at users and their access. After all, the system would be meaningless without its users. It's like a city without residents or Coca-Cola without bubbles. Certain elements are indispensable. So, in this chapter, we'll take you on a wonderful journey through user management and all the elements related to their access.

I would like you to pay special attention in this chapter to how you can grant specific access to your users, and how to revoke it – you could say that you are somewhat the lord and master of your Salesforce org. Here is a list of topics you will soon explore:

- User management
- Password management
- Controlling system access
- Profiles, roles, and permission sets
- Sharing settings and organization-wide defaults
- Field-level security
- Login policies and **multi-factor authentication (MFA)**

User management

Do you remember how Dr. Frankenstein created his monster? Just like him, we will create our first user (don't worry; it will be much less macabre than in the novel). During your work in this ecosystem, you will often encounter requests to create a new user. When setting up a new user, the request should include certain significant elements such as the following:

- First/last name
- Email address
- Username
- License
- Profile

These elements are extremely important; without them, users can't be created and in the later stage access the system. Creating a new user can be compared to setting up your own account in a new online store. Just as there, you need to provide details such as first/last name, email address, and phone number. Often, systems also ask for a username.

And here, let's pause for a moment. I want to draw your attention to one important element. In Salesforce, your username must be unique, meaning it can't appear in the database twice. Can the username be the same as the email address? Of course. But does it have to be the same? No. You can freely manage the usernames of your users. If you are tasked with creating an entirely new org and managing it, then consider your structure of usernames. Think about whether you want the usernames to be the same as email addresses, or just "first name.last name," or maybe the first letter of the first name, the full last name, and the company domain. The choice is yours.

> **Tip**
>
> If you work for a consulting company that is a Salesforce partner and you, as an employee, have the opportunity to collaborate with many companies on their orgs, I encourage you to use the method of adding the client's domain to your credentials. For example, if you work for XYZ company and your email address is john.doe@xyz.com, and you start collaborating with FunnySocks, ask to set up a user for you with your email address and the username john.doe@funnysocks.com. In the future, this will make it easier for you to manage your login data, as the domain at the end will let you know which username is assigned to a particular client.

If we already have some knowledge about the necessary elements, I'll show you how to create a new user step by step.

As with most administrative actions, we need to go to our favorite setup under the gear icon in the top-right corner. You surely know it very well.

1. Next, as shown in the following screenshot, in the quick find box, write users and click on **Users**:

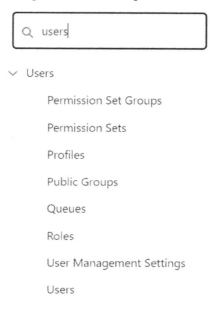

Figure 6.1 – Quick Find box and Users

2. On your screen, you will see a list of all available users. At the top of the table, you'll find three buttons: **New User**, **Reset Password(s)**, and **Add Multiple Users**. We are interested in the first one. The other two options are the ability to reset the password for one or more users and to add multiple users at once:

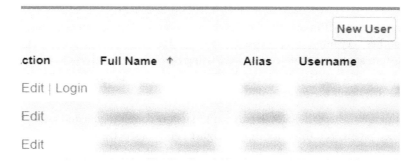

Figure 6.2 – New User button

In front of you is a sizeable form with quite a few fields to fill out. Let's go through the most important ones. As I mentioned earlier, here we have, among others, a name, email, username, user license, and profile. All these elements should be provided by the person making the request. During configuration, we can also add attributes to the user such as **Knowledge User** (having access to the knowledge base from Service Cloud) or a user with access to Service Cloud:

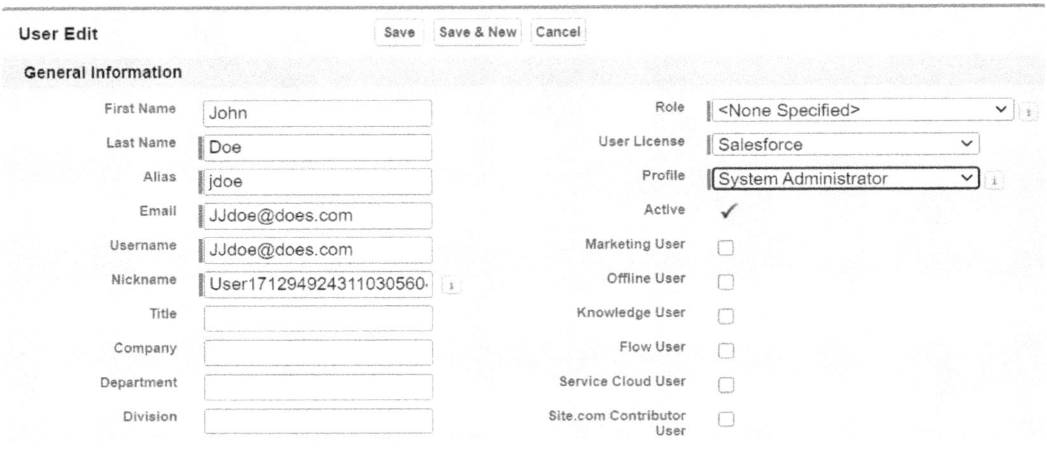

Figure 6.3 – User details

After entering the data, selecting the appropriate role, license, and profile, scroll to the bottom where you'll find the **Generate new password and notify user immediately** checkbox. Unchecking this will stop the system from sending the details to the new user. Why is this needed? One example is to grant access to all users at one moment. Once you click the **Save** button, you have a fresh user in the system.

However, I'd like to draw your attention to a very important element during user creation: the license. Salesforce offers a plethora of licenses with very diverse functions and accesses. One of the most powerful, akin to *"One Ring to rule them all"* from *The Lord of the Rings*, is the Salesforce license. This is what you get in the starter package and what Salesforce sellers pitch. It gives you access to all objects in a given org, and, with the right profiles or permission sets, you can perform many actions in the system. Another very interesting license is Salesforce Platform, primarily designed for creating and deploying custom applications/objects and accessing them. It allows access to any custom object created in the Salesforce ecosystem. So, if you have users who only use custom objects, confidently assign them to the Salesforce platform.

Now, imagine a situation where the sales department manager comes to you and asks you to remove John Doe from the system. You think it's simple, right? Let me surprise you: you can't delete users. I was once surprised myself; wanting to clean up the user database, I looked for the **Delete** button and couldn't find it. I figured there must be a way to delete the user. I entered the Developer Console, but I couldn't do it there either. Then, someone much wiser than me explained that there's a pretty important reason for this. It's historical data; to maintain data continuity, we can't delete users because some records would lose their owners, and historical data would not be integral.

So, what can you do? Deactivate! Simple, right? Well, there are two options for revoking a user's access to the system: deactivation and freeze. But each has its specific application; let me list them for you:

- Deactivation, editing the user, and unchecking the **Active** checkbox completely revoke the user's access. This is most commonly used in case of an employee leaving the company or changing positions and no longer needing access to Salesforce. In this case, the license returns to the pool of available licenses.

- Freeze is a temporary block on account access. We do this by going to the user and clicking the **Freeze** button. This scenario is used if there's some temporary proceeding against the employee, and they might return any day. In this case, the license does not return to the pool of available licenses.

As you can see, there are many options when creating a new user, so it is good to understand their needs and access requirements. Remember that access to Salesforce for new users is like a well-tailored suit – if it fits well, you'll use it often, but if it's uncomfortable and doesn't sit right, you'll opt for a high-five with your favorite sweatpants and skip the outings that require dressing up.

Password management

This section can be summarized by two characteristics of passwords: hard to break and containing as many different types of characters as possible. But you probably already know that passwords such as 12345 or QWERTY are not good choices. In general, it seems that the silliest passwords, such as 1L1k3P4nK3ke$ (I like pancakes), are the hardest to crack. Unfortunately, we don't always have control over the passwords our users set... or do we? Here and now, I want to make you aware that we can force them into certain password requirements. To do this, go to **Setup**, and in the quick find box, please enter password policies and select the entry that appears. Here is a list of settings that will help you manage passwords using **Password Policies**:

- **User passwords expire in**: This setting allows us to determine the length of time a password is set for. In the list, we have options ranging from **30 days** to **Never expires**.

- **Enforce password history** (How I dislike this setting for my own passwords): This prevents the reuse of the last passwords, from "No password remembered" to even 24 passwords. So, "ILovePancaces24" would still be valid.

- **Minimum password length**: The minimum length of a password. Generally, the longer, the better, but it can also be easier to make mistakes.

- **Password complexity requirement**: This setting determines the difficulty of passwords in the system.

- **Password question requirement**: Ensures the password is not found in the security question generated during a password recovery attempt.

- **Maximum invalid login attempts**: The number of incorrect password attempts during a user's login attempt.

- **Lockout effective period**: The time a user must wait after entering the maximum number of incorrect passwords during a login attempt.

- **Obscure Secret Answer for Password Resets**: An option that hides the correctness of the answer when responding to a security question.

- **Require a minimum 1-day password lifetime**: Ensures that the password must be used for at least 1 day before changing it again. This prevents too frequent password changes, which could potentially lead to a security breach.

- **Allow use of setPassword() API for self-resets**: The use of the `setPassword()` API for resetting passwords. A feature typically for developers.

- **Forgot Password / Locked Account Assistance**: This section allows us to set up a custom message in case of forgotten passwords or account lockouts.

- **API Only User Settings - Alternative Home Page**: An option for users who only have API interface access. A user with such access, when attempting to reset their password, will be redirected to a specified URL; that is, the alternative home page.

Salesforce offers the possibility of personalizing many settings, and it hasn't disappointed us in the case of passwords either. We can set specific password policies for a given profile. Why might this be useful to us? For example, we can set more restrictive password rules for administrative employees due to their access to the system, while Standard users can have shorter and less complicated passwords.

Resetting passwords is one of the more important duties of an administrator. Unfortunately, I'm not joking. You will often encounter emails, phone calls, or tickets with requests for an immediate reset. And what can you do about it? Simply reset the password. Being in **Setup | Users**, you can reset the password on the user's record or the user list. Another way is to educate users on how to do it themselves. This education will take longer, but it will pay off. Teach your users that on the login page, they have the option to reset their password by answering their secret questions. I recommend this from my own experience.

And with that, we have exhausted the topic of password management in Salesforce. This option gives a lot of power, and we, as administrators, must act according to company policies, but remember our users – those 24 remembered passwords might not make you many friends among them.

Password Policies is a powerful tool that assists in establishing secure access to your Salesforce org. However, that's not the only action you can take. In addition to **Password Policies**, you can also implement additional levels of security to prevent unauthorized access. One of these features is associated with IP login ranges. Let's explore the specifics of this feature in the next section.

Controlling system access

Once we have users and they have their passwords, in theory, we can say that the system can start to live. Using the earlier example of a city and its residents, our users can start wandering through our system, visiting its nooks and crannies, creating records, and admiring dashboards. But they also need to sometimes leave this digital city and relax amid the real chirping of birds, the sound of water in the nearest stream, and the warm rays of the sun on their skin... Oh, forgive me – I got lost in thought looking out the window at the snowflakes swirling in the wind. But returning to the topic of mandatory time off from Salesforce, many companies have their policies, a set of rules that apply to employees at a certain level.

Sometimes companies simply do not want employees to access Salesforce data outside of the workplace, which may be due to data security reasons or simply because the office is located in a certain place and the company does not offer the possibility of remote work (such a scenario often occurs in various types of call centers or help desks). Salesforce has enabled the setting of a specific range of IPs from which users can access the system. So, leaving work, you cannot log in to your account because the IP is different. This is also an easy way to secure our system from unwanted guests and filter out unknown IPs. To set up IP, in **Setup | Quick Find**, type network access and click on **Network access**. You will get a list of already entered IPs and the possibility to add a new IP. The form to enter a new IP looks like the one in *Figure 6.4*:

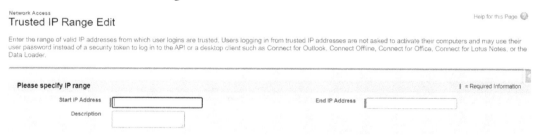

Figure 6.4 – New IP range form

An important aspect of setting the IP range for login is that it does not completely block access to the system. If a traveling salesperson or someone working remotely tries to log in, the system will ask them for a verification code, which will be sent to their email. However, if the company has a stationary call center, where employees can only work from a specific location, it is worth considering setting such an IP range at the profile level. This setting will prevent receiving a verification code and logging in from anywhere other than the workplace.

If we are talking about call centers, many such departments have their specific days and hours of operation. So, if the call center operates on weekdays from 8:00 A.M. to 6:00 P.M., why not configure this?

To do this, you need to make a certain change in the user profile. Let us do this together:

1. Where do you think we will start? Yes! **Setup**, then enter `profiles` in **Quick Find** and click on the searched location.

    Select the profile you want to modify. In my case, it will be **Standard User**.

2. After clicking, you will see the profile settings. I ask you to scroll down to the bottom and find **Login Hours**. If it was not set before, you will see a blank space with **No login hours specified**.

    Click **Edit** and enter the hours for the appropriate days, as shown in *Figure 6.5*:

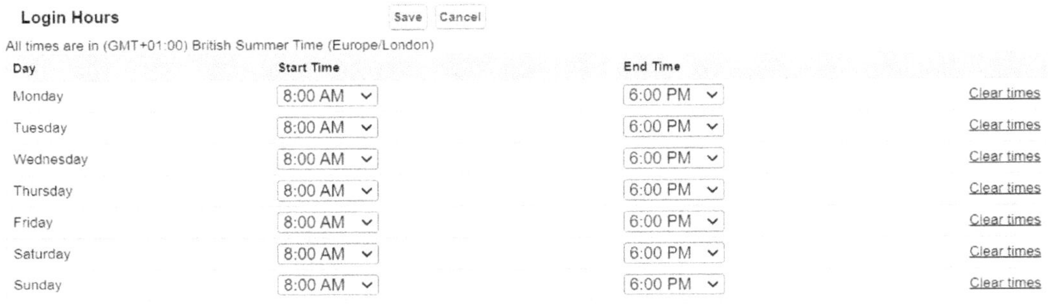

Login Hours		Save	Cancel	
All times are in (GMT+01:00) British Summer Time (Europe/London)				
Day	**Start Time**		**End Time**	
Monday	8:00 AM ⌄		6:00 PM ⌄	Clear times
Tuesday	8:00 AM ⌄		6:00 PM ⌄	Clear times
Wednesday	8:00 AM ⌄		6:00 PM ⌄	Clear times
Thursday	8:00 AM ⌄		6:00 PM ⌄	Clear times
Friday	8:00 AM ⌄		6:00 PM ⌄	Clear times
Saturday	8:00 AM ⌄		6:00 PM ⌄	Clear times
Sunday	8:00 AM ⌄		6:00 PM ⌄	Clear times

Figure 6.5 – Login Hours

You probably noticed that I made a silly mistake with Saturday and Sunday. But I assure you that I did it intentionally. By entering the same time in **Start Time** and **End Time**, we prevent users from logging in for 24 hours on that day.

Once we have set the days and hours, we click **Save**, and there we have it! Our login hours for the call center are set. Remember that all logged-in sessions after 6:00 P.M. will be logged out, and users will have to log in with their credentials the next day or after the weekend.

The last and crucial element of access management to the system is **Session Settings**. The most important setting here is **Session Timeout - Timeout Value**. Here, we can set the duration of the session without logging out in Salesforce. Why is this option extremely important? When a user logs in to the system at an internet cafe and does not log out, the system will do it for them. Okay – I'm joking; who uses internet cafes these days? Only secret agents. But seriously, not everyone is in the habit of locking their computer when leaving their workstation. In my environment, such situations are used to send messages such as "Pizza on me" on the group chat on Slack. But not everyone may have such good intentions and will want to snoop in Salesforce. Therefore, to prevent such situations, we can introduce a session length. This time can be set from 15 minutes to 24 hours of inactivity. A reasonable time is about 2 hours of inactivity. It will allow returning to the workstation after lunch and getting back to unsaved records without fear of data loss.

The last point I want to mention in this section is that you can apply the preceding settings for specific profiles; you do not have to set the same settings for everyone. So, if we want to make life easier for our boss, remember that those 24 hours of inactivity for them will be quite a nice gift.

Profiles, roles, and permission sets

If we already have a user, let us give them access to certain doors that were previously closed to them. Therefore, in this section, we will deal with profiles, roles, and permission sets. Let me start with the first one. Long, long ago, Salesforce created great software for database management. When users started browsing all the records, one of them stumbled upon the salaries of other employees. This is, of course, a fictional example, but I think such situations could have been quite common. Whenever I start a conversation with a client, I mention such examples; they always spark the imagination, and we start discussing access.

Profiles in Salesforce are a key element in ensuring user access and security. Forgive me for spoiling a later part of the text – this may soon change. For now, profiles maintain all elements related to access; thanks to them, you can grant or revoke the possibility of entering records of a given object. Can there be a user without a profile? No – there cannot be; every time, we must choose a profile when creating a new user. It is profiles that define our access to objects, fields, tabs, or functions in Salesforce.

There are two types of profiles:

- Standard profiles are those provided by Salesforce in the out-of-the-box version. They have specific names and predefined settings. Most often, they are the basis for our custom profiles, which we can create ourselves. Believe me – setting up access on a clean profile is a long-hours job for very persistent people. Among the standard profiles, you will find such stars as our favorite **System Administrator**, **Standard User**, which has access to (as the name suggests) standard objects, **Read Only**, following the principle "you can watch but you cannot touch," and many others. Next time you log in to the system, go to **Profiles** and see their full range.

- Custom profiles are for special tasks. Sometimes we have a group of users who have specific access. To illustrate this better, imagine the company JustCoin, which deals with trading gold coins. They have a portal that allows the purchase of coins online. But as with any portal, sometimes something can break, or a customer may not read all the steps in the instructions, and unfortunately, the "turn it off and on again" rule does not work. So, at JustCoin, there can be two types of problems – one from the technical side and the other from the purchasing side. Fortunately, as is the case in different companies, there is a help desk that can assist customers. The system admin, while creating the Salesforce platform, specifically Service Cloud, set different access for these two teams. One dealing with technical problems got access to accounts, contacts, cases, and the knowledge base – so they could efficiently resolve their customers' issues. The other team, the one assisting with the purchase of gold coins, got access to Account, Contact, Opportunity, Products, and Pricebooks, so they could provide good advice to their customers in commercial matters. If both teams had the same access, their performance could be lower, and the technical team could peek at customers' prices and purchases, which could affect data security. So, if you need to tailor something to the needs of your company, it is best to clone the profile closest to your needs and adjust it to what your users expect.

Creating profiles is quite a lengthy process. However, if you are tempted to do so, let me lead you to the place where you can forge your first profile. To create it, do the following:

1. I will not surprise you. Setup. Yes – we enter our good old friend **Setup**.

    Then, we type `profiles` in the **Quick Find** box and click on the item that comes up.

    Now, a long list of profiles available in your org should be in front of you.

2. If you want to create a new profile, click on **New Profile**:

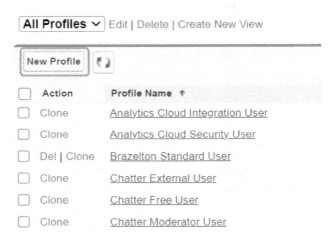

Figure 6.6 – New Profile button

And here, Salesforce helps us because it allows you to choose a profile to be cloned. It is based on this profile that you will be building your new one. After selecting it, all that remains is to give it a name and save it.

Now, you have your new profile. To modify it, just click on it, enter the section you are interested in, and click **Edit**. Here, you can start modifying access.

After making changes, save them.

Tip

To avoid unnecessary security breaches, regularly check profiles and review their accesses. With tools such as Advanced Salesforce Inspector, you can easily check individual settings. Ensure that you act cautiously when assigning profiles. Remember that each incorrectly assigned profile grants specific visibility, and sometimes it is not intended for a user with the wrongly assigned profile.

The next element we will discuss is roles. It sounds like I am about to cast you as Macbeth in the latest adaptation of Shakespeare's work. But it is not that – roles in Salesforce are used to define organizational hierarchy. That has an important impact on access and data visibility in the system. By building a hierarchy in Salesforce, we can reflect what it looks like in the real world, in the organization for which we are configuring the system.

Therefore, when assigning such roles, try to describe them with standard CEO, CIO, CFO, CIA, FBI – okay, not the last two, but the first three will certainly be welcome in the role structure. Roles determine what data users can see in the hierarchy. Usually, the hierarchy structure allows for the visibility of records assigned to the user and records owned by users located below them in the hierarchy. To better visualize this, look at the following screenshot:

Figure 6.7 – Roles hierarchy

See – CEO and CFO are the standards. So, if you look at the example in the screenshot, you can be sure that the CEO will see what the Director of Channel Sales sees, but the SVP of Human Resources will not see the records of the COO.

As you can see, using roles can be a fantastic way to establish visibility for a certain structure within an organization.

Now, we move on to the third of the great titans of access – permission sets. Essentially, a permission set grants access. But how are they different from profiles? Well, they are more detailed. Suppose you are a system admin in a large IT company; only you and one other person have access to the physical server in the entire company. So, despite your passes, which allow you to move around the entire building, you will also receive an incredibly special card that will allow you to enter the server room. The same is true for permsets (that's a nickname – I recommend it), which are certain passes to specific settings. They allow you not to have to make changes globally for the entire profile. Suppose you work in a nonprofit organization. This organization helps troubled youth. You collaborate with psychologists who have their database in the Salesforce Nonprofit Cloud. They collect all sensitive data and enter it into the system, obtaining results for psychoanalysis – addiction risks. The team of psychologists has access to contacts but also to a custom object collecting this data. So, you have prepared a special "Psychologist" profile for them. But there is also a user who coordinates new therapy participants and needs to know how many meetings there were and what the results were, to help them even better. At the same time, they must have access to Accounts, Contacts, custom referrals, and many other objects, but most importantly, access to the data collection object. They, like other office employees, have the **Office team** profile. To give this one employee access, you would have to create a new profile. However, it is much easier to create a new permission set that will grant them access to the meeting data object. Easy? Sure – it is easy.

With permission sets, you grant individual access without globally changing profiles. Now that you know what these individual accesses are, you will learn how to create such a setting:

1. Click on **Setup** under the gear icon, and in the Quick Find box, type permission sets.

2. After clicking on this item, a list of available permission sets will open. There, you can also clone a permission set or create a new one.

   If you want a completely fresh permset, click on the **New** button.

3. Assign a name (label), API name, and description.

   Set **Session Activation Required** – when this option is enabled, the user must activate the session before gaining access to the permissions specified in this permission set. This means that the user must additionally confirm their identity (for example, through 2FA) to access the additional permissions contained in the permission set. This is useful when you want to increase the security of access to sensitive data or functions.

   Then, choose the license on which the permission set will operate.

4. In the next step, choose the accesses and save the changes.

An interesting and quite new solution is permission set groups. This solution allows you to create groups of permission sets. In other words, we can create a permission set that allows access to specific objects and builds accessibility for certain user groups with them.

A crucial element is still the ability to grant appropriate permissions in access – so-called **CRED** permissions. It is an acronym for **Create, Read, Edit, and Delete**. You can grant these accesses in permission sets as well as in profiles.

Do you remember, as I mentioned at the beginning of this section, I am going to spoil it a little? Let me get to that topic. Quite recently, Salesforce announced that it will slowly phase out the functionality of profiles, and instead, will use permission sets. It is important to note that profiles will remain but will hold less data; only the following will remain:

- Login hours/IP ranges
- Record types and app assignments
- Page layout assignments (old ones, not App Builder with Dynamic Forms)

And the rest of the permissions would go to permission sets. Salesforce dates this change to 2026. So, if you are reading this book in 2026, permission sets have already dominated Salesforce, and AI has not created Skynet, so we are safe (I hope so).

In the next section, I will introduce you to the secrets of sharing settings in org-wide defaults. I warmly invite you to it.

Sharing settings and organization-wide defaults

In the previous section of this chapter, we learned how Salesforce profiles, roles, and permission sets can influence Salesforce security related to Salesforce objects and their features. In this section, we will learn about security related to Salesforce data sharing. To do this, we will deep dive into the world feature called **Sharing Settings** to discover its two core features called Organization-Wide Default (often shortened to OWD) and Sharing Rules. Both features are connected and together create the core on which the Salesforce security is built. Let's see how those features work in detail.

Organization-Wide Default

From the previous section of this paragraph, we understood that Salesforce profiles and permission sets are responsible for giving access to Salesforce objects. So, profiles and permission sets are controlling this if I see the **Lead** tab, **Account** tab, **Opportunity** tab, or any other standard or custom tab. But which Salesforce feature controls which data I should see in objects I gained access to via profiles or permission sets? The answer is Organization-Wide Default. OWD is a Salesforce feature where the Salesforce administrator can decide if the data from certain Salesforce objects should be visible to the users. OWD helps you set the default organization access to the data related to each object. It enables you to decide who besides the record owners should see the records and edit them. To access OWD, just search `sharing settings` in Salesforce **Setup**. Let's see what options we have when setting up the default sharing setting for Salesforce objects.

Types of object data access in OWD are the following:

- **Private** – Only records owners and people above them in the Role hierarchy have access to them and can edit them.

- **Public Read Only** – All Salesforce users have access to all records, but they can only view them not edit them. However, the record owner and those higher in the Role hierarchy can modify the records.

- **Public Read/Write** – All Salesforce users have access to records and are able to edit them. Here, it does not matter if you are higher in the Role hierarchy or not as everyone in your Salesforce org can see and edit records not owned by them.

- **Public Read/Write/Transfer** – Used on Leads and Cases. Besides viewing all records with the possibility to edit them, you will be also able to transfer them to other users.

- **Controlled by Parent** – Used mostly when the object is a child in the Salesforce master-detail relationship where is set to default. It can be also used on some Salesforce standard objects such as Contacts, Orders, or Assets, which gives you the option to base the object's records visibility on the object that is parent to Contacts, Orders, or Assets.

When setting the default internal access, you will also see the **Grand Access Using Hierarchies** checkbox on the user interface. This small feature is very important as thanks to it, users above the record owners in the Salesforce Role hierarchy can have access to their records the same way as the owners (can view and edit records).

Besides **Default Internal Access**, you can also set **Default External Access**, which is access to the records when showing them to the Salesforce Experience Cloud users. **Default External Access** must be more restrictive or equal to **Default Internal Access**.

There is, of course, a strong connection to profiles' or permission sets settings because even when the Salesforce object OWD is set to **Public Read/Write** but the user doesn't have edit permission granted via a profile or permission set, they will not be able to edit the record. Moreover, there are magical permissions that are available on profiles and permission sets that when added can upgrade the record's visibility of certain users on objects even where the OWD is set to private. They are named **View All** and **Modify All**. When you have **View All** permission related to a certain Salesforce object, you will be able to view all records stored on this object despite any OWD setting. Similarly, when having **Modify All**, you will be able to see and edit all records stored on this Salesforce object. The Salesforce world can be complicated, I know!

Sharing rules

Okay – but what if the OWD settings and Role hierarchy are not enough and you would like to give some users more rights toward some object's records? You can of course modify the OWD setting but the changes made there will influence all Salesforce users. To add some more rights to the object's, records you need to use the **Sharing Rules** feature.

Sharing Rules can open access to some records to some set of users. How does it work? Instead of talking more about theory, let's show how it works in practice. Let's create one sharing rule together.

We will be creating a lead sharing rule that will give Channel Sales Team members access to Leads owned by other team members. This rule makes sense when the Lead object OWD is set to private so that only record owners can see their own leads (plus people above them in the Role hierarchy) and for some reason, you would like to give all team members view access to leads owned by other users. Follow this step to create a lead sharing rule on your Salesforce team:

1. Go to **Sharing Settings** in **Salesforce Setup**.

2. Click **New** in the **Lead Sharing Rules** section.

3. For Set Label, choose **Channel Sales Team Leads**.

4. Select your rule type: **Based on Record Owner**.

5. Select which records are to be shared by going to **Roles| Channel Sales Team**.

6. Select users to share with by going to **Roles | Channel Sales Team**.

7. Select the level access for the users: **Read Only**.

After the sharing rule creation, you will see the **Lead Sharing Rules** section, as shown in the following screenshot:

Figure 6.8 – Lead sharing rule example

Okay – now that we have done the simple exercise, let's see and explain the other options we could use when setting a Salesforce sharing rule:

-   **Rule Types** options – there are two types available. The chosen type affects the next section, **Select which records to be shared**:

    -   **Based on record owner** – The sharing will be based on the ownership of the records. For example, you can share records owned by members of some public groups or roles or roles and subordinates with other groups of users. There is no possibility of choosing a single user when setting this kind of sharing.

    -   **Based on criteria** – The sharing will be based on records field criteria. For example, you can share only Leads on some specific status or Accounts with type equals competitor, and so on.

- **Select which records to be shared** – Depending on the rule type, the sharing can be based on the group or criteria, so basically records field values:

  - Group options:

    - **Public Groups** – The owner is part of chosen Salesforce public groups.

    - **Roles** – The record owner has a certain role assigned; for example, Sales Director.

    - **Roles and Subordinates** – The record owner has a certain role assigned or is a subordinate of this role. So, for example, if the Sales Director's role has Sales Representatives as its subordinates, then in this case the Sales Representatives will be a part of this rule.

  - Criteria options:

    - Choose the field, operator, and value to create a records filter.

    - Set the filter logic if needed.

- **Select the users to share with** – Similar to the **Selected to be shared** group options:

  - **Public Groups**

  - **Roles**

  - **Roles and Subordinates**

- Record access options – The final setting that will control how records mentioned in the **Select which records to be shared** option will be visible to users mentioned in the **Select the users to share with** option:

  - **Read only** – Users will be able to see the records mentioned in the rule. As you can see, the first option here is **Read Only** not **Private** as **Private** is always the option from which we would like to give some users some more access.

  - **Read/write** – Users will be able to see and edit records mentioned in the rule.

Tip

As you're aware, the visibility of records is primarily governed by the ownership of the record but also strongly by the Role hierarchy. This implies that users positioned higher in the Role hierarchy will inherit record access from those beneath them in the hierarchy. Because the hierarchical sharing is already done by Salesforce Roles, in most cases, Sharing Rules are used when you need to horizontally share record visibility; for instance, among teams or groups situated on different branches of the Salesforce Role tree. A common example involves sharing access to certain records between two teams reporting to different managers. This scenario frequently occurs in customer support, where teams may need to view cases from other teams to provide support when one team is overwhelmed by the number of cases.

Okay – now that you understand how Salesforce OWD and Sharing Rules function and their impact on record visibility among users, let's delve deeper into controlling field visibility on those records. We'll explore this in the next section.

Field-level security

Field-level security is a straightforward concept that enhances security beyond object and record access. Put simply, it regulates field visibility for users. By employing field-level security, you can dictate whether users with particular Salesforce profiles or permission sets should have access to specific fields or not.

Let's see how this works in practice. Let's modify the field access on the **Account** object:

1. Navigate to **Setup** and search for **Field Accessibility**.
2. Choose the **Account object**.
3. Choose the **View** by **Fields** option.
4. Choose a field. For example, pick the Phone field
5. Choose the profile that you want to update; for example, **Read Only**. Click the link in the **Field Access** column.

    Edit the **Field-Level Security** settings. Let's make the **Phone** field not visible for the user with the **Read Only** profile. Just make sure that the **Visible** checkbox is unchecked.

6. Save your settings.

Please look at the following screenshot. You should have a similar outcome:

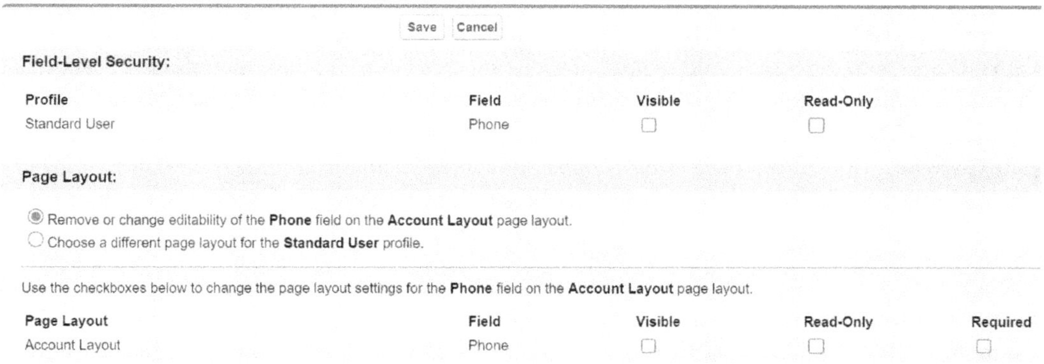

Figure 6.9 – Field-level security example

Voilà! You've successfully hidden the **Phone** field for the user with the **Read Only** profile. Easy-peasy! We could conclude this section at this point, but let's delve into a few more details concerning field-level security.

Here are some additional important points:

- You can hide fields and set them as read-only

- You can also adjust the **Page Layout** settings concerning the fields managed via field-level security. This means that using field-level security, you can perform the following settings on the field: hide, set as read-only, or make it mandatory. The changes you made will be applied in the **Page Layout** setup.

- Making a field hidden or read-only at the **Profile** level holds more significance than doing so at the **Page Layout** level. For instance, when a field is hidden due to a page layout, users with the corresponding profile still have access to the field via Salesforce Reports. Conversely, if the field is hidden via **Profile** on the field-level security, it will not be visible through Salesforce Reports

- You can also review or update the field-level security for the profile. Simply select **View by Profile** when configuring field accessibility.

Utilizing field-level security is straightforward yet significantly impacts system security and the user interface. You've also learned that concealing fields or setting them as read-only at the **Profile** level holds much more strength than implementing the same actions at the **Page Layout** level.

You now understand how to set up Salesforce OWD, establish sharing settings, and utilize field-level security to effectively share data among users is a pivotal responsibility of a Salesforce administrator. This authority comes with great responsibility, ensuring that users only access the intended data. However, there's no need to worry once you've acquired the knowledge to configure these features.

Login policies and MFA

In the previous section, we discovered how to secure the data internally so that the proper user who has already access to your org could see the proper data and not more than they should. Now, let's see how to secure access to the org itself. In this section, we will handle the security topics related to Salesforce user's MFA. This topic is very important, and each company should be aware of it. As Stéphane Nappo, Cisco Security Officer, once said: "*It takes 20 years to build a reputation and a few minutes of cyber incident to ruin it.*" Let's now see how Salesforce creates a secure org environment using MFA.

MFA stands as a straightforward yet highly efficient method to fortify login security, offering robust protection for your business and data against potential security threats. MFA involves a process necessitating users to confirm their identity through two or more verification steps before gaining access to their Salesforce account. Several methods can be used to secure your org. Let's list and further explain them in detail:

- **Salesforce authentication app**: The easiest, most convenient, and cost-free among all the MFA options in Salesforce. It's a Salesforce native mobile app for iOS and Android that will be used by your users to be able to log in to your org.

- **Third-party authentication app**: Third-party authenticator applications capable of generating **time-based one-time password (TOTP)** codes. Works similarly to the Salesforce authentication app but it's not a Salesforce product. Among the popular options, we can list Google Authenticator™ and Authy™.

- **Security keys**: These can be physical devices that use some public-key cryptography. I know that it looks a bit old-school, but some companies are using this kind of physical key, and Salesforce gives you the option to use it too.

- **Built-in authenticators**: Verify identity with a fingerprint, iris, or facial recognition scan, or a PIN or password.

As you can see, the Salesforce native app is the most recommended option when implementing MFA in Salesforce. Of course, other options are also valid but are often related to additional costs, and their implementation is more complicated and time-consuming.

Now that we know which authentication option we have, let's see what the MFA implementation options in Salesforce are.

Starting from February 1, 2022, applications developed on the Salesforce platform mandate MFA for user access to your org's user interface. You have the option to enable MFA universally for all users at once or implement it gradually in phases for specific user groups who log in directly using a username and password. Possible implementation options are the following:

- **Enable MFA for your entire org** – You have the option to activate MFA organization-wide using a single configuration. Once enabled, every internal user accessing their accounts with their username and password will be required to provide a secondary verification method.

- **Enable MFA for specific users** – You have the flexibility to begin implementing MFA by initiating a pilot program or gradually introducing it to your users through phased rollouts using the MFA user permission. Once MFA is activated, users logging in to your organization directly with their username and password will be required to present a secondary verification method.

- **Exclude exempt users from MFA** – certain scenarios are exempt from the mandatory MFA requirement. As Salesforce implements and enforces MFA in the future, many of these scenarios will be automatically excluded. However, there are a few cases where customers need to exempt themselves. If any of these situations are relevant to your environment, utilize the **Waive Multi-Factor Authentication for Exempt Users** user permission prior to the activation of MFA for your organization, either by your own initiative or by Salesforce in future updates. Avoid assigning this permission to any internal users accessing your Salesforce org's UI, which encompasses admins, privileged users, standard users, developers, as well as users authorized to represent your company, such as partners and third-party agencies.

Controlling logins and setting up MFA is one of the most crucial tasks for a Salesforce administrator. In the age of cybercrimes, you need to function as a cyber guardian to safeguard your data and users from potential internet threats. Now that you're aware of the Salesforce features that can enhance your org's safety, we also recommend further studying these topics in Salesforce Trailheads. As software is a living organism, staying updated with Salesforce's best practices is essential to remain informed about any new features that can enhance or alter organizational safety.

Summary

In this chapter, users delved into pivotal elements of Salesforce security and access control.

The exploration began with an in-depth look at sharing settings, shedding light on the crucial role played by organization-wide defaults and sharing rules. Users gained a profound understanding of how these settings dictate record access within an organization. Particularly emphasized was their role in horizontal data sharing among different teams or groups, enriching the comprehension of data visibility.

Field-level security emerged as a critical aspect, unveiling its significance in managing field visibility and editability. Users discerned the nuanced differences between controlling field access at the **Profile** level versus the **Page Layout** level, empowering them to exercise tighter control over sensitive data.

The chapter progressed to focus on login policies and MFA, exploring strategies to fortify user authentication and ensure secure access to Salesforce platforms. This segment provided insights into implementing robust login policies and the integration of MFA, strengthening the overall security infrastructure.

Regarding Salesforce security, the motto "*This is my house, I have to defend it*" (Kevin, *Home Alone*, the movie) mirrors Salesforce admins' commitment to fortify its protection. For this reason, throughout this insightful journey, you have garnered a comprehensive understanding of fundamental Salesforce security measures. Controlling who can see records, managing what specific information people can view, and making sure only authorized users can log in are very important for keeping data safe and making sure the right people can use Salesforce properly. When you know how to protect your Salesforce system let's now see how to automate things. We will familiarize ourselves with this topic in our next chapter.

7

Automation Tools

In the previous chapter, we delved into the intricacies of user management. I hope that this will help you manage your users. This chapter is where the fun starts, as it will introduce you to automation. Sounds interesting? I think so.

In this chapter, we will cover topics that are essential in the daily life of your organization, namely the following:

- Approval processes
- Flow Builder
- Apex triggers
- AI in Salesforce

Approval processes

We'll start this section with a question: do you remember as a kid asking your parents for permission to have sweets or to spend time playing Warcraft or Starcraft with friends? I think all of us asked our parents for something at some point. But what if such a request was automated? Imagine your parents/caretakers receiving an automatic query the moment you reach for the first M&M. Or imagine you start your favorite game and it pauses until you receive approval from a parent. Sounds like a nightmare for kids, but it's a perfect solution for parents. This, in a nutshell, is what an approval process is. With the high dynamics of current times, users working on Salesforce may forget certain steps or simply not adhere to existing procedures. But we, the system admins, stand guard.

To help you further understand what an approval process is, let's describe a real scenario. Imagine you are the manager of a large sales department and the end of the year is approaching, so all the salespeople in your team want to hit their targets. Therefore, they sell a lot very quickly. If they want to sell a lot, they need to activate higher discounts. Sounds logical, right? But we must remember that the company needs to make a profit. So, if a salesperson gives too high of a discount, the company's profits will be negligible. Here comes the knight in shining armor—the system admin!

You ask them to set up a certain business process for you. It will be based on the amount of discount given. To help you understand this scenario, I'll remind you of what the sales department might work on—opportunities. It's on this object that they will have the main sum of sold products and a field for entering the given discount. The salesperson is the owner of the opportunity, so the management knows exactly who is managing this sale.

As a sales department manager, you have already contacted your system administrator, and with their help, you have created three approval processes:

- **≤ 30% discount**: The opportunity is locked during the approval process. A request for discount approval is sent to senior sales, and you only receive a notification about the assignment of such a discount. After acceptance or rejection, the opportunity is unlocked.

- **30%–40% discount**: The opportunity is locked during the approval process. A request for discount approval is sent directly to you. After acceptance or rejection, the opportunity is unlocked.

- **40% discount**: The opportunity is locked during the approval process. A request for discount approval is sent to the **chief revenue officer** (**CRO**), and you receive a notification about the initiation of such an approval process. After acceptance or rejection, the opportunity is unlocked.

I think the next step in our IT journey will be to show you how to create such a process.

We will create it together step by step. But first, go to the Setup, and in the **Quick Find** box, find **Chatter** and click on the result. There, find **Allow users to receive approval requests as posts** and set it to **True**. It will permit you to add the **Approval Post Template**, which is an approval in the chatter (hell yeah!). Now, I will show you what the approval process creation form looks like. You can find fields such as **Name**, **Email template**, and **Fields** that will be used as conditions. It is shown in the following screenshot:

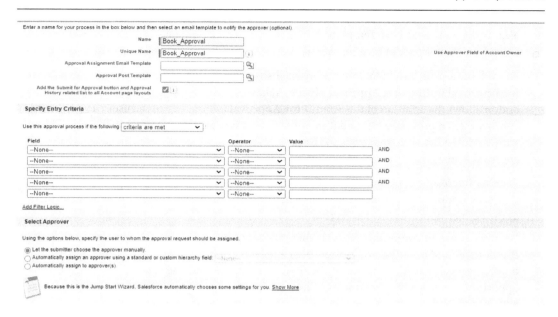

Figure 7.1: Approval process form

And now, a step-by-step manual.

1. Open **Setup**.

    I.   In **Quick Find**, type **Approval Process** and click on the found item.

    II. Select the object you intend to work on from the **Manage Approval Processes For** list.

    III. After selecting, you will see a list of previously created approval processes divided into **Active** and **Inactive**.

2. The next step probably won't surprise you; to create a new approval process, click on the button labeled **Create New Approval Process**.

    I.   Here we get two options: **Use Jump Start Wizard** and **Use Standard Setup Wizard**. These are two options for creating this process. Try both and see which one works better for you. I chose **Use Jump Start Wizard**.

    II. A new page will open where you must enter the name of the approval process, and from this, a **Unique name** will also be created.

    III. Then, you need to select an **Approval Assignment Email Template**, which will be the email template sent when an approval is assigned to a user.

3. Next is the **Approval Post Template**, which, like the email template, is optional. In this case, it will be a chatter post informing the user in the assigned process.

    You will see a small checkbox with a quite extensive **Description**, which reads **Add the Submit for Approval button** and **Approval History** related list to all opportunity page layouts. This option allows the addition of the **Submit for Approval** button, where the user can personally approve such a process without automation. Such a button can be added to the page layout with a related list with approvals to the page layout. This allows you to review and see the approval history.

4. Another interesting checkbox on this form is the **Use Approver Field of Opportunity Owner**. This is an option to create a custom field where a user can choose another user and use them as the defined approver for that record.

5. Next, we move to conditions. This is where the real fun begins. In this step, you can set the appropriate conditions that will activate your approval process. For example, on the opportunity object, you can use a condition of the discount amount and the appropriate stage for the opportunity, e.g., a 40% discount and a negotiation/review stage). Then, the approval process will be triggered.

    If we're talking about conditions, remember that you don't have to rely only on fields and the values available there. You can also use formulas. When setting conditions through a formula, check its logic first using the **Check Syntax** button.

6. In the last section, you have three options for approvers. The first allows the person initiating the approval process to choose the approver. The second is the approver from a given field. The third is the default selected approvers.

    After choosing all these options, all that remains is to confirm all the settings using the **Save** button. But is that all? I'll disappoint you: no. Welcome to the next step.

    We have already defined the basic elements of the approval process. But there are still steps that will be taken during such a triggered process. The first of these is **Initial Submission Actions**, i.e., actions that will be triggered during the launch of the process. The main action, which is set by default, is to lock all changes to the record. You can add a change here, e.g., you can change the status from **Negotiation** to **In Approval Process**.

    The second is approval steps, i.e., actions that will be taken after each approval. In the case of several approvers, this is very useful information. If you click on **Show Actions** at the first default step, you will see that there are two types of actions—those that are triggered in the case of acceptance and those that are triggered in the case of rejection of the request. We can create automatic tasks, field changes, outbound messages, and email alerts for them.

    The next section is **Final Approval Actions**, i.e., what will happen when the record is accepted by all approvers. In it, by default, we have created the action **Record Lock**. Yes, it was already in the previous one, but if the record has already been accepted in the form it was in, changes are not advisable.

The penultimate section is the evil twin brother of **Final Approval Actions**, called **Final Rejection Actions**. This is a very similar section, with a small difference—this action unlocks the record so that the owner can make the necessary changes and do an approval recall.

The last section is **Recall Actions**, i.e., what exactly will happen when the record is recalled.

If you want to polish your approval process, click on edit at the top of the form and select **Specify Approver Field** and **Record Editability Properties**. There, you can configure elements such as who the next approver should be chosen by and options for who can edit records when they are locked for other users. We have an advantage here because only administrators can make such changes.

7. For the last step, we're going to, as they say, "Get the London look", or "Pimp My Approval Layout". Yes, here you can add fields that will be displayed to approvers, providing them with the necessary amount of information to help them make good decisions.

I'm proud of you! You've made it through one of the longer configuration processes in Salesforce. I know it may seem complicated now but trust me, if you configure two or three such processes, you'll be a master at them.

> **Tip**
> Remember to test each of your processes before deployment. Even the smallest error can cause issues, such as locking a record. So, create and test a record from an opportunity or another object, and even before QA steps in, go through all the scenarios you can think of.

Approval processes are hugely significant for companies and their internal policies. If they are well-designed and updated, they serve as a solid defense against unprofitable deals or bad decisions. An approval process can be combined with Flow, the native automation in Salesforce, which you will learn about in the next section of this chapter.

Flow Builder

Modern times could be compared to jumping into hyperspace from Star Wars (sorry, Star Trek). If we close our eyes for a moment, we might feel like Steve Rogers (Captain America) after 66 years of being frozen. Sometimes I wonder what Walt Disney would think after such a long freeze, but he would probably like the movie Frozen (ba dum tsss). But getting back to the topic, we often do not notice the conveniences that surround us, from mobile phones to streaming platforms and many other impressive elements of our daily lives. One such improvement is Flow in Salesforce. It is automation—a process that is automatically triggered based on certain conditions, waiting for specific behaviors or a specific time/day. What was before Flow? Salesforce employees monitored each record and when conditions were met, they made the required changes. Of course, I am joking; workflow rules came first, and then there was Process Builder. The former was a form where you set conditions and behaviors. The latter introduced a visual interface for mapping automation. It was much more powerful than its predecessor, as it could create records, launch Flow, invoke Apex code, and much more. It allowed for multiple IF/ THEN conditions. An example process looks like that in the following figure:

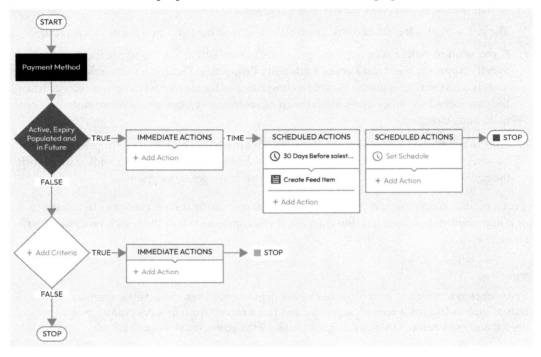

Figure 7.2: What does the Process Builder look like?

Next came Flow. It revolutionized automation in Salesforce. I remember when I first encountered it. I thought "*Wow, this is fantastic.*" And right after that, I thought "*Oh no, I have to learn this from scratch.*" But it was a pleasant learning experience of discovering something new.

What is Flow? It is a powerful automation solution that allows users to create automated processes without writing code. This tool enables the creation of very simple automation, such as sending birthday wishes to a user when it is their birthday. Other automations are more complex, such as a multi-branch automation that checks all users, creates a list from them, appends the appropriate permission sets, and adds TRUE to a hidden checkbox to generate a report.

There are many benefits to using Flow in Salesforce. Certainly, one is increased operational efficiency, i.e., automating repetitive processes and eliminating the risk of human errors. Personalization of business processes involves adapting technology to business requirements, making the solution better suited to corporate and client needs. An essential element is also the scalability of the project; in this case, every flow can be changed, expanded, or reduced according to our needs. Using Flow also translates into improved data quality. Thanks to the repeatability and availability of only a group of data, automation using it standardizes all records.

Salesforce, like any business, has ensured that you have a choice. What kind of flow do you wish for? There are six types of flows; let me take you through them:

- **Screen flow**: This type allows you to create an interactive user interface based on a specific process. It is most commonly used for collecting customer data. Often, during insurance purchases, for example, the seller must enter necessary data in a predetermined sequence. Thanks to this sequence, the seller knows they have recorded your name and surname, and also the mileage of your car.

- **Auto-launched flow** (**no trigger**): This is automation triggered by other processes or events in the system (including Apex or REST API). No user interface is required here. For example, a large software development company wants to streamline the flow of completed software versions for testing. Once developers mark a version as ready for testing, an Apex or REST API activates a flow to generate a series of test tasks for the QA team and sends them a notification stating that the version is ready for testing.

- **Schedule-triggered flow**: This is automation triggered according to a selected schedule (daily or weekly). During configuration, you can also choose the time when the flow is to be launched. For example, a charity wants to remember its donors and send them emails on their birthdays. This flow will scan users, create a list with the condition `Birthday date = TODAY`, and send them a prepared personalized birthday email.

- **Record-triggered flow**: This flow lies in wait for changes to records. It is triggered by specific actions such as create, edit, or delete. To this, we can add conditions that the record must meet to be considered. For example, you want to ensure that users who want to opt out of receiving emails from your company will have their data anonymized. Every time the opt-out checkbox on the contact record is true, our flow starts and replaces the data, such as first name, last name, email, phone, or address, with a pre-selected sequence of characters.

- **Platform event**: This flow activates upon receiving messages from platform event notifications. For example, if a phone or app reports an error, a message is sent through Apex, other flows, or processes. With the received message in Salesforce, the system generates a new case with user data and error ID.

- **Record-triggered orchestration**: This type uses orchestration, i.e., the coordination of several complex actions related to the Salesforce database.

  For example, a telecommunications company receiving a report from a customer can assign the case to a service agent, schedule a repair date and time with a technician, notify the customer, and, in the meantime, initiate a problem diagnosis.

As you can see, Salesforce has a wide range of automation and everyone can find something for themselves. But with flows, it is a bit like with Lego bricks—you might have a place to lay them out, but what do you do when you do not have the bricks themselves? Here come the key components of the flow, riding in on a white horse. Let me introduce them to you:

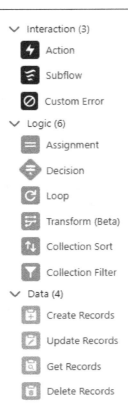

Figure 7.3: All flow components

Now I'd like you to take a look at all the Flow components you may encounter when building a new automation:

- **Action**: This is a special agent for tasks. Seriously though, it's used for assigning tasks, sending emails, triggering approvals, and much more. For example, you can send an email notification to a customer after updating the case assigned to them.

- **Subflow**: This is a bit like flow-ception—using another flow within a flow.

   For example, during the purchasing process, a subflow is triggered to verify the correctness of the CVV number of the credit card.

- **Custom error**: This component handles custom errors. For example, an error might read, "*Oops, the CVV number seems to be from another world, not this card. Check it and enter the correct one.*"

- **Assignment**: This involves assigning different values to variables;

   For example, assigning the discount amount depending on the customer's status and the size of the order.

- **Decision**: This directs the flow in specific directions based on previously configured conditions. For example, when a new customer makes a purchase, they receive a welcome email with a starter discount, while a regular customer receives a thank-you email and a loyalty discount.

- **Loop**: This is a very popular element in coding. It is an iterative processing of data sets.

  For example, it gathers all people who have birthdays today. The loop searches all contact records, retrieves all the necessary data, and, using an email template, sends birthday wishes from the company.

- **Transform**: This involves transforming data in sets. This could be changing data format, structure, or values. For example, you can change the address structure and transfer data from a contact to a new address object.

- **Collection sort**: This helps in sorting data sets.

  Example: Sorting the list of orders by order date.

- **Collection filter**: This filter collected data sets, e.g., filtering client records by the region of residence.

- **Create records**: We get God mode and can automatically create records in the system. For example, you can create new records of a custom object meeting after a customer expresses interest in joining the company's VIP club.

- **Update records**: This updates existing records in the system. For example, it updates case records after a technician has completed repairs on the day of their visit to the customer.

- **Get records**: This retrieves record data, e.g., warehouse status to check product availability.

- **Delete records**: This handles the deletion of records from the system.

  For example, you can delete outdated leads from the system.

I would very much like to tell you that this is all there is to know about Flow. But I will not do that because we still have resources to cover—elements that store or manipulate data. There are many types of resources in SF, such as variables, constants, formulas, or collection choice sets. **Variables** store data, much like a folder in which information is collected and can be used or changed once closed. **Constants** stand their ground and do not budge; they hold unchangeable values. **Formulas**, something we have already encountered, perform calculations. Collection choice sets are used to store larger amounts of records or values and are extremely useful when working with loops.

To start any adventure with Flow, familiarize yourself with its types, components, and resources, which will help you in later designing automation. Once you know what each thing is, it is worth understanding the business goals at the start. If it is your idea, ask yourself the most important question: what do you want to achieve by using this flow? A clear understanding of the goal is helpful. Then, map out such a process and create a visual map of it, e.g., a chart. Finally, choose the right type of flow. With these three elements, you are (almost) destined for success.

When building with flows, as with system configuration, we follow only the best practices. Thanks to the SF community, we can learn about them and test them ourselves to see if they work. One of the important best practices is modularity and the reuse of flows. If you build automation that is triggered during changes to an account, and in the future, you need the same flow but with a different action, just use the old one. There is no need to build 10k flows for one object. Another best practice is to anticipate errors, test, and validate. This way, you avoid emails from users with subject lines such as "Weird message on contact." Thanks to the debug option, you can check your freshly created structure and plan tests.

As I mentioned earlier, Flow is a bit like building with blocks, starting from the main piece of your puzzle and expanding it with additional modules. In the following figure, you can see a fairly simple flow containing decision elements and **Create Records** element:

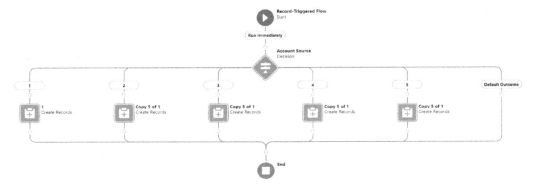

Figure 7.4: Simple flow

You might ask yourself, "Will I be able to do this?" I am happy to answer that. In Flow, principles of logic and the basics of programming are used, such as the loops I mentioned earlier. Of course, understanding them will not be as demanding as passing a course and an exam in Java in Mandarin. I assure you that after a few configurations, you will be able to do them without any problem, and you will be able to answer what a loop is, even if you just woke up at three in the morning.

> **Tip**
>
> If you want to learn Flow, I encourage you to visit the website `https://salesforceflowlab.wordpress.com/`. It's a treasure trove of knowledge based on real use cases. The exercises will guide you step by step on how to perform a given flow. In the end, they'll spice things up by adding a few additional requirements to your previously built flow.

Okay, we have our flow, and we are very proud of it and want to save it. Let's show the world our creation... But after pressing the **Save** button, a big red message appears saying that something is wrong, just like in the following screenshot:

Figure 7.5: Error while saving the flow

Is that bad? Of course not. Thanks to this message from SF, you can verify your errors and make the necessary corrections. Every time you try to save, the system checks the correctness of your configuration, and if something is wrong, it will eagerly notify you. But our favorite system would not be itself if it did not offer us an additional tool for checking our configuration. I have mentioned it before: debug. When you click on this magical button, a new form will open, just like in this next screenshot:

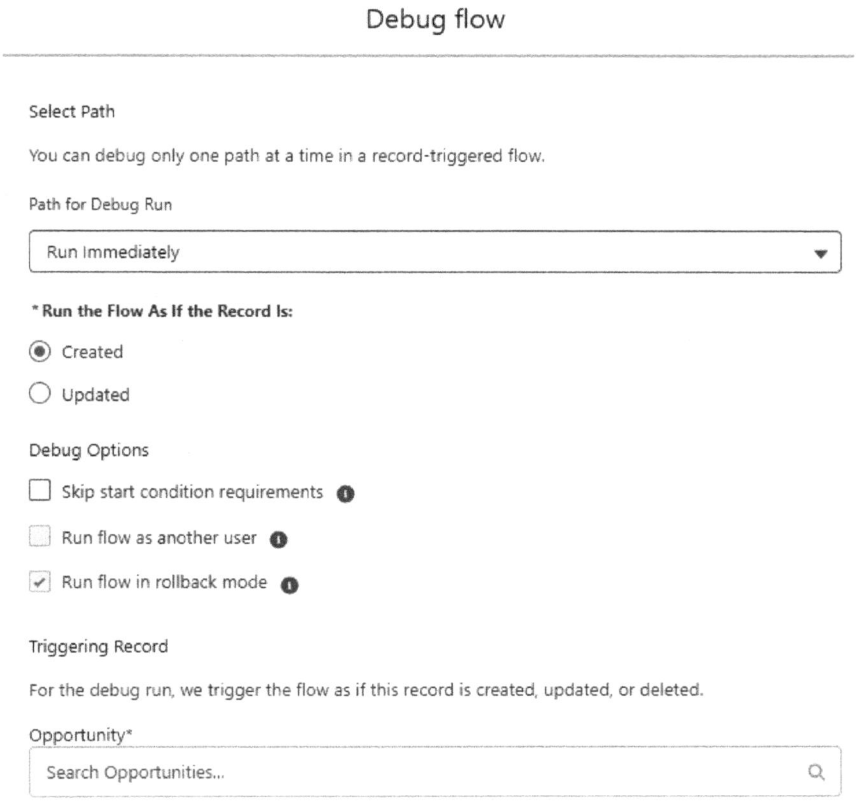

Debug flow

Select Path

You can debug only one path at a time in a record-triggered flow.

Path for Debug Run

Run Immediately ▼

* **Run the Flow As If the Record Is:**

◉ Created

◯ Updated

Debug Options

☐ Skip start condition requirements ❶

☐ Run flow as another user ❶

☑ Run flow in rollback mode ❶

Triggering Record

For the debug run, we trigger the flow as if this record is created, updated, or deleted.

Opportunity*

Search Opportunities... 🔍

Figure 7.6: Debug form

On it, we have a few options that will allow us to test our automation.

The first, **Path for Debug Run**, is a tool that tracks and analyzes the exact path of our flow, following every step. This allows you to see which elements of the automation were activated and at which point our flow derailed.

Skip start conditions requirements bypass the main conditions for starting the flow, meaning that if we set that opportunity to be in the Prospect status, and we want to test our flow on our test opportunity, we use this option so that the initial conditions are not checked.

To see how our creation works on other users, we can use the run flow as another user option. The run flow in the rollback mode option restores the previous state of the record, meaning changes made in it will be reversed.

- We can run flow in two methods—created or updated
- The last selection field is choosing the record, where we can select our test record for the trial

Do not forget about documentation. It will help you remember in the future what this automation was executed for and its purpose. Carefully describe each element of your configuration; later, if anything breaks, you will be able to find the elements that caused the errors. Finally, describe what the flow is responsible for. Your successors and colleagues will thank you.

When should we even think about automation? I think the users of the system you will be managing can answer this question. If you want to help them, arrange meetings with different departments, and make sure that the meeting includes a manager/lead as well as people in lower positions. These are the people who often perform repetitive tasks, which can somehow be simplified and automated for them. Do not ask them, "What can we automate for you?" Instead, ask, "Imagine a system without limitations and tell me what would make your life easier. What would be useful if the system could do it for you without any effort?" These users will be able to tell you what can be improved in their processes and daily work.

As I mentioned at the beginning, the world is moving forward. Automations are everywhere and help us in many aspects of our lives. Remember, to utilize flows 100%, carefully prepare for what business goals you have to achieve, plan your project, and start building using the best practices.

Modularity, testing, validation, and error fixing are key during the construction of automation. These are elements that will be extremely useful in building automation that will work for us for years. But is there something else? Something more advanced? Yes! Apex! More on that in a moment.

Apex triggers

Programming in Salesforce can serve many purposes, and one of them is creating automation to support users' work within the application. Mastering Salesforce programming requires a separate book, and although writing a simple Apex trigger isn't complicated, it goes beyond the scope of this book, as it isn't solely dedicated to programming in Salesforce. Nonetheless, we'll briefly mention here what Apex triggers can be used for and how they differ from declarative automation such as flows. We'll also provide an example of a simple Apex trigger that could be written to automate processes in Salesforce.

Let's begin by defining what an Apex trigger is precisely. An Apex trigger is a code typewritten in the Apex programming language. Apex triggers allow you to execute personalized operations before or after modifications are made to Salesforce data, such as inserts, edits, or deletes.

There are two types of Salesforce Apex triggers:

*   **Before triggers** are utilized to modify or authenticate data values of a record before they are saved in the database.

*   **After triggers** are utilized to modify or authenticate data values of a record after they are saved in the database. For example, you will use the after trigger when there is a need to access data such as **RecordId** or **LastModifiedDate**, as some of the field values will be available only when the record is saved to the database.

Salesforce Apex triggers can perform actions such as creating, editing, or deleting records, and those actions can be executed before or after the following types of operations:

*   `insert`
*   `update`
*   `delete`
*   `merge`
*   `upsert`
*   `undelete`

Now that you know what actions an Apex trigger can perform and when the trigger can be activated, let's examine what an actual trigger may look like. Check out the following code snippet showcasing Apex trigger syntax:

```
trigger TriggerName on ObjectName (trigger_events) {
                code_block
                }
```

If you wanted to write an Apex trigger that operates on the `Lead` object and fires before inserting a record into the database, here's an example of how such a trigger invocation would appear:

```
trigger myLeadTrigger on Lead (before insert) {
    // Your code here
}
```

Of course, this APEX code does nothing, as it lacks any executable content. Let's add a few lines of simple code to change it into the real deal:

```
trigger LeadDescriptionTrigger on Lead (before insert) {
    for(Lead l : Trigger.new) {
        l.Description = 'Your new Lead description';
    }
}
```

As you may have noticed, this code performs a very simple task. Before each new lead is saved to the database (before insert), it enters the value Your new Lead description into the **Description** field on the lead record. This isn't necessarily the best business application for Apex triggers, but it certainly gives you an idea of what an Apex trigger looks like and when it can be applied. You might now be thinking, "Why do I need this whole Apex trigger when I can create automation based on flows?" Yes, you're right. Many, if not most, business scenarios related to automation can be handled using Salesforce Flow. However, if you require efficient examination of intricate procedures in batch situations, then the adaptability of Apex, along with its extensive debugging and tooling features, is the ideal choice for you. The complex logic that can be used in Salesforce triggers involves the following:

• Defining and evaluating complicated logical expressions or formulas

• Complex list processing, loading and transforming data from large numbers of records, and looping over loops of loops

• Anything that requires map-like or set-like functionalities

Additionally, implementing an Apex trigger is not as straightforward as Salesforce flows because Apex triggers cannot be created and activated directly in the production environment. To introduce an Apex trigger to the production environment, you need to first create it in the sandbox, develop a specific Apex test class ensuring sufficient code coverage, and only then deploy it to production. Salesforce flows, on the other hand, can be constructed and activated directly in the production environment, which makes it riskier when someone decides to implement them in this manner. Therefore, while Apex triggers are considered more deliberate and secure, it's important to note that everything is relative, and a poorly written Apex trigger can be as risky for your data as a poorly written Flow. Due to the continuous and rapid evolution of Salesforce Flows, they are increasingly taking on the roles and tasks that were previously fulfilled solely by Apex triggers. The addition of before flows, scheduled flows, and other features has resulted in flows becoming the primary choice for many automation-related tasks instead of Apex triggers.

Now that you've learned how to create Salesforce automation by yourself, let's explore how you can be relieved from these tasks and discover what Salesforce supported by artificial intelligence can do for you!

Salesforce AI

The previous year was a turning point in the widespread adoption of artificial intelligence (AI), showcasing its presence across various domains. AI's versatility was evident, not only as a text and image generator reminiscent of renowned artists such as Picasso, Goya, or Banksy but also in its substantive role within business applications, notably Salesforce. This maturing technology transcends creative realms, demonstrating its prowess in aiding data analysis, thereby enabling more accurate and insightful decision-making processes.

Salesforce, recognizing the transformative potential of AI, strategically integrated AI functionalities into core segments such as sales, service, marketing, and e-commerce, as well as Data Cloud. This integration extended the impact of AI beyond mere creativity, fostering predictive analytics capabilities that significantly augmented operational efficiency and personalized customer experiences. Let's delve into the details of what Salesforce had to offer in late 2023 regarding AI-supported features.

Einstein for Sales

Salesforce's most popular cloud received several features supported by AI. The AI helps the user to get the best information to write emails and notes and contact your prospects. See in detail what Salesforce AI has to offer in the Sales Cloud:

- **Automated research assistant**: The Salesforce AI makes prospect and account research automatic. It will gather important information from various web sources directly into your CRM and update current records automatically as you discover new details. Just like with ChatGPT, you can communicate with Salesforce AI with proper prompts and ask Einstein about data summaries, specific information about accounts, leads, contacts, and opportunities, and other things.

- **Sales emails**: This feature will let you quickly create personalized emails using CRM information. This will help salespeople introduce themselves, set up meetings, or remind contacts about follow-ups in just seconds. AI will automate personalized messages using Salesforce and other data sources, available in Microsoft Outlook, Gmail, or LinkedIn—wherever you're working.

- **Einstein Copilot for sales and call summaries**: Thanks to AI, you can forget typing and writing notes. Salesforce will create brief, useful summaries of your sales calls. It can recognize key points, customers' feelings, and future actions to assist the sales team in advancing deals. You can modify summaries and easily share them via Slack or email, promoting collaboration across teams for better deals.

- AI also will help sales teams move forward by telling reps what to do next, keeping everyone focused on the most important tasks for better productivity.

- Thanks to this feature, you will be able to identify important information such as customer concerns, competitors, and pricing details in your one-on-one conversations and support your decision with it. You can easily share these summaries in Slack or email for better teamwork and deals.

- **Buyer assistant and Einstein relationship insights**: This feature will help to change your static lead forms on your website to live chats to move your leads to sales quicker. You will be able to decide faster which leads are good and send them to salespeople from your site. Interested leads will talk to a salesperson right away and will be able to pick a meeting time that works for them.

- **AI-driven deal insights**: Thanks to deals insights, Salesforce AI will help you focus on the most important deals, understand them, and act faster on them to create healthy opportunities and perform other key sales actions.

- **Activity data sync and contact data creation**: AI will automatically gather and sync important customer and sales details from your email and calendar.

- **AI predictive forecasting**: Salesforce will improve guessing accuracy with smart AI predictions. You will be also able to understand the reasons and trends used for these predictions. AI will highlight the gaps and problems in your forecasts.

As evident, Einstein for Sales brings a lot to the table, and that's just the beginning. Let's now explore how Salesforce AI can assist your customer support teams.

Einstein for Service

Salesforce's second most popular product also gained numerous AI-supported functionalities. This is particularly important because customer service departments can now operate even faster and more effectively. AI supports communication with your customers, writes articles for your Salesforce knowledge base, categorizes for you, and proposes the next steps related to resolving customer issues. The following are the key AI functionalities associated with Service Cloud:

- **Conversation summaries and service replies**: Salesforce AI will help agents work faster and make customers happier by automatically analyzing conversations making smart summaries and then creating AI-generated replies on SMS, WhatsApp, and other platforms. With Einstein Service Replies you will quickly understand customer chats and will get AI-created answers from these chats or your company's Salesforce knowledge base. Your support agents will be able to easily send these replies to customers or change them before sending.

- **Knowledge articles creation and usage**: Salesforce AI will support your users with articles made after conversations with your prospects and customers. Besides automated article creation, Salesforce will also assist agents and customers in finding answers quicker with AI-driven knowledge searches. It will show instant answers directly on the search page or agent console.

- **Case classification, next best actions, and replay recommendations**: Salesforce Einstein looks at past cases, fills in new case details, sorts them out, and sends them to the right person. It helps agents finish work faster and get it right.

Besides case analysis, Salesforce can support your users by suggesting next steps and replay recommendations.

Einstein Bots: Thanks to this feature, you will be able to create a bot that works across many channels and languages, linking to your Salesforce data easily. This AI solution will boost your team's productivity by helping with everyday tasks and questions.

Einstein for Sales and Einstein for Support are the core AI functionalities that Salesforce currently provides. But that's not all! Let's explore what else Salesforce has to offer!

Einstein for Marketing

If Salesforce Sales Cloud and Service Cloud gained AI support, there couldn't be a lack of such support for marketing efforts on the Salesforce platform as well.

Thanks to journey orchestration, multichannel messaging, and built-in AI support, you will be able to create and send customized customer journeys to increase sales interactions to deliver great marketing campaigns and content marketing that can be used in email, mobile push, SMS, third-party messaging, or advertising.

Einstein for Commerce

E-commerce is one of the fastest-growing sectors in today's market, so AI support in Salesforce Commerce Cloud was a necessity. What's more, the beneficiaries of such support won't be limited to just you and your users; it will include practically every one of your customers. AI not only assists you in configuring products but also effectively suggests them to your customers. Let's explore the functionalities of AI that will enhance Commerce Cloud:

- **Commerce concierge and Einstein product recommendations**: Salesforce AI will make your customers' shopping experiences better with AI-powered personal shopping help. Commerce Concierge works with messaging apps, giving easy-to-understand help using words, pictures, and data. It helps find products, gives personalized answers, and makes buying online easier for a smooth shopping experience. At the same time, Salesforce Einstein tracks shopping trends to suggest popular items to your customers. It also gives recommendations based on how shoppers use your online store.

- **Generative product descriptions**: Thanks to the Einstein GPT, you will save time by using AI-generated, personalized product descriptions.

- **AI-supported promotion management**: Salesforce Einstein GPT will support you with auto-generated custom offers for email, SMS, and messaging apps in multiple languages.

- **AI-driven chatbots**: Those multilingual AI connect it with your customer and product details so users can easily explore, find information, ask questions, and buy things.

As you can see, there is a lot that Salesforce AI can do. It supports sales, customer service, marketing, and e-commerce in everyday tasks and duties. All of this generates data that humans want to analyze to make better future decisions. Can AI help us with this as well? Yes, it can! It does so with Einstein for Data.

Einstein for Data

The most recent tool benefiting from AI support is Salesforce Tableau. With intelligent predictions and recommendations, it guides actions to enhance sales, customer service, and related areas. Tableau's AI-driven insights offer strategic advice, aiding in sales enhancement, customer service optimization, and other crucial business facets.

Einstein Copilot

At the time of writing this book, just one month before its completion, Salesforce announced its latest AI-related product, Einstein Copilot. Our knowledge of this tool is based solely on the Salesforce press release, where we glimpsed at some screenshots and saw a few demos. So, what exactly is Einstein Copilot? According to Salesforce, "It is seamlessly integrated across Salesforce applications, providing a consistent user experience that can answer questions, generate content, and dynamically automate actions for enhanced productivity, deeper customer relationships, and higher margins".

Einstein Copilot's reasoning engine interprets intent and chooses optimal actions, functioning as a processing mechanism for informed decision-making, problem-solving, and insight generation. Notably, Copilot is built into your Salesforce organization, ensuring precise content alignment with your data and processes. It seamlessly integrates into the Salesforce UI, offering users quick access to efficient work. Think of it as ChatGPT for Salesforce, tailored to your specific data. No need to worry about data privacy, as this is secured within Salesforce's dedicated Einstein Trust Layer.

Now, how does Copilot work, and what are its use cases? Can we converse with our Salesforce? It may sound like sci-fi, but Salesforce's demonstrations make it seem like you're interacting with your CRM system! For instance, a customer service agent can ask Einstein Copilot to close a case, open a sales opportunity, or sell an add-on. The copilot will comprehend the user's intent and execute tasks seamlessly within the service experience. There are no silos between applications or data. Einstein Copilot can summarize your Salesforce data, suggest actions, complete the actions, do some analytics, and more. Einstein Copilot somehow centralizes the way you interact with all the AI capabilities mentioned in this chapter, so it can support sales and speak with sales representatives about their records and life plans (just a joke, guys). At the same time, it can help support users and their managers, the managers of their managers, and so on. What's intriguing is that Salesforce admins and developers can extend Copilot with new actions to meet specific needs.

As you can see, AI-related functionalities are developing so rapidly that by the time you read this book, they will likely be even more advanced than what we've just described. Let's just hope that no all-powerful machine takes over the world, even if it comes from a cool tool such as Salesforce. Otherwise, we might find ourselves singing songs in its honor as a punishment for not updating and taking care of it properly. It's, of course, a bit of an abstract joke, but who knows? I've had a saying ever since AI started surrounding us: "*I always use the polite word 'please' when asking something from AI; you never know who will be in charge soon!*"

Summary

In this chapter, you've learned many practical aspects related to automating processes in Salesforce. We discussed core features in Salesforce, including approval processes, flows, Apex Triggers, and automation using AI.

With the knowledge of approval processes, you will be able to independently create automation that allows you to build approval processes for your business with a wide range of applications.

In the section about *Salesforce flows*, we broke down the entire flows feature into its fundamental components. You learned the purpose of flows, the types of flows, and the actions that can be triggered in the system through flows. Flows are now the primary automation tool available in Salesforce, and Salesforce is actively developing this feature. By familiarizing yourself with flows, you will become a low-code Salesforce developer.

In the *Apex triggers* section, you learned about the different kinds of triggers in Salesforce. These triggers are snippets of code that perform actions before or after specific events, such as adding, editing, or deleting records.

You discovered examples illustrating how these triggers are built and how they can automate tasks or execute specific actions based on events within Salesforce.

In the *Salesforce AI* section, you explored Salesforce AI by learning about Einstein and its uses in sales, service, marketing, and e-commerce, as well as Data Cloud and Einstein Copilot.

You have discovered how Salesforce's Einstein AI empowers sales, support, marketing, and e-commerce with predictive analytics, records insights, engages with customers, makes recommendations, and understands data.

The rapid development of artificial intelligence tools will undoubtedly lead to their increasing use in Salesforce, far beyond their current applications. How will this look in a year or two? Certainly, it will be exciting and different, but "*The future has not been written. There is no fate but what we make for ourselves*" (a quote from *The Terminator*, 1984). Meanwhile, before AI rules the software world, or perhaps the entire world, just before Judgment Day, let's revisit learning how to do things with our own hands and manually execute some tasks related to Salesforce reporting. Join us in the next chapter to explore *Salesforce Reports and Dashboards*.

Reports and Dashboards

In the previous chapters, we learned how to create and automate data in Salesforce. All right, but what comes next with all this data? Do they exist in the CRM just for the sake of being there? Is Salesforce like a boring novel that someone created and nobody wants to read? Of course not! At least it should not be like this! There will come a moment when you'll want to report on the data that have made their way into your CRM to support both your administrative activities and the actions of your users, their supervisors, and even your organization's CEO. As you already know, Salesforce is a highly sophisticated tool that doesn't overlook reporting modules, and they are quite advanced. In this chapter, we will break down the components of Salesforce reports and dashboards, giving you valuable insights that will help you and your colleagues analyze data to make the most informed business decisions.

In this chapter, we will cover the following topics:

- Creating reports
- Dashboards

Creating reports

Salesforce is a CRM software that stores data for everyday use. Users are inputting this data, updating this data, or deleting the data. An everyday work routine is built from those tasks, and all of this data stored in Salesforce can be reported and exported using Salesforce reporting features, which can be used in several situations, such as the following:

- Tracking users' actions, such as opportunities, tasks, cases, etc., in reports
- Creating the reports to, in turn, build Salesforce dashboards
- Exporting the data outside Salesforce

Salesforce reporting features are quite extensive. However, before you learn more about how to create Salesforce reports, let's now see how to create report folders in which you will be able to store your reports.

Report folders

Salesforce report folders are similar to Windows folders. Similarly, you can just create the folder and place the report in it. Let's see how to do this step by step:

1. Create a report folder, as shown in the following image:

Figure 8.1: Creating report folder

2. Name the folder. For example, if you are storing sales-related reports, you can name them Sales Reports. If you use it to store data related to Salesforce cases, you may like to name it Support Reports, as shown in the following image:

Figure 8.2: Naming the report folde

3. Give users access to the folder: After creating the folder, you need to give your user proper access to the folder. Click **Created by Me** along the left side, under **Folders**. Then click into your new Sales Reports folder. You will use the **Share** option to accomplish this task. You can find the **Share** option after clicking the "arrow" button in the **Salesforce Report** tab, as shown in the following screenshot:

Figure 8.3: Sharing the report with users

When adding user access, you have three options:

- **View**: The user will be able to view reports in the folder
- **Edit**: The user will be able to view and edit reports in the folder
- **Manage**: The user will be able to view, edit reports, and manage access to the report folder to give other users access to it

In the following image, I have shown you what needs to be chosen to manage folder access:

Share folder

These sharing settings apply to all subfolders in this folder.

Share With

Users

Names		Access
Krzysztof Nowacki ✕		View
		✔ View
Share		Edit
		Manage

∨ Who Can Access

Done

Figure 8.4: Choosing the report sharing Access type

While sharing the reports with users, don't forget to click the **Share** button to confirm access.

1. Create subfolders: If needed, you can create a report subfolder. Just enter the already created report folder and create the next one inside it.

2. Save the report/s in a folder: Your report folder is ready to be used. You can now save the reports inside, and users will have access to the folder and will be able to see or edit them.

As you can see, creating Salesforce report folders and subfolders and giving users access to them is very straightforward.

> **Important note**
> Salesforce serves you with some predefined report folders. You can use Private Reports and Public Report folders to store the reports. A Private Reports folder is only accessible to you, whereas the Public Report folder is accessible to all users.

Now that you know how to create a folder and where you can save all your reports, let's build some reports!

Report builder

Okay. So you are already prepared and have the proper folder or folders. Now let's see how to become the report master and create advance Salesforce reports.

Let's see how to create a report:

1. Create a report

2. Choose a report type

3. Save the report

Is this so simple? Yes. And no! To quickly create a report, you do not need to do more than this. When choosing the report type, the report will be served with some predefined report columns and basic filters, so if this is what you are looking for, then your report journey will be fast and easy-peasy. However, to be honest, I have never used a report with predefined columns and filters. It will always be something that you would like to change, whether this is adding new report columns, changing report filters, summarizing the data, or highlighting the most important details. Let's break down all the most important report features.

Report columns

You can manage the report columns in the **Outline** section. Just search and pick the field that you want to add as a column. The field will be displayed last, but you can rearrange the fields. Just drag and drop any field available in **Outline**.

After adding the field to the reports, you are able to do the following:

- **Sort Ascending**

- **Sort Descending**

- **Group Rows by The Field**

- **Bucket the Column**

- **Show Unique Count**

- **Move Left**

- **Move Right**

- **Remove Column**

See the following image on how to find the field options mentioned in the preceding list:

Figure 8.5: Report column options

Now that you know how to deal with Salesforce report columns, let's learn how to use report filters.

Report filters

Report filters give you control over the data visible in the report. Although it's a main feature of Salesforce reporting, it's a very simple element. The easiest way to learn about Salesforce report filters, is "learning by doing," so let's see how to add a simple filter to a Salesforce report to see only the customer accounts.

1. Go to the **Filters** subtab

2. Search for the Type field and mark the types of accounts that you would like to see in the report: **Customer – Channel** and **Customer – Direct**:

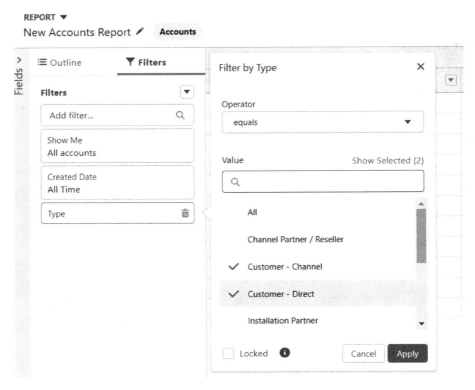

Figure 8.6: Report filters example

Press **Apply** and **Run** the report to see the report results. On the same screen, you will see the records you were looking for and the filters you applied. The following is an example of a run report:

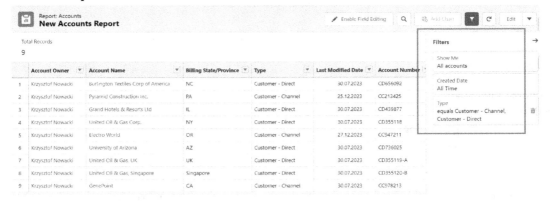

Figure 8.7: Final report with filters

We have just used a report filter related to the picklist field, but based on the field type, the filters can have different options. Let's now see how the filters would look if you would like to filter by data or data/time fields. Let's try to filter using **Create Date field** and see the filter options. When using date or data/time fields, in the filters, you are able to choose certain dates or relative dates. When choosing **Use relative date**, you can use sentences such as This Week, Last Week, Last Quarter, Next Week, etc., instead of a precise date. See the date filters example in the following image:

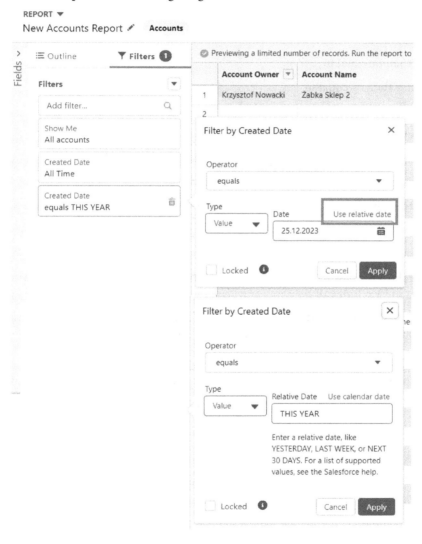

Figure 8.8: Report date filter usage

As you can see, using report filters is very easy and straightforward. You can use any field that exists on the object on top of which you are creating your report. As you can see, when creating an account-type report, you can set the filter operator and create filters. See the operators in the following list. There are many of them, but their usage is very intuitive and easy:

- **Equals**

- **Not equal to**

- **Less than**

- **Greater than**

- **Less than or equal to**

- **Greater than or equal to**

- **Starts with**

- **Ends with**

- **Contains**

- **Includes**

- **Includes all**

- **Excludes**

- **Excludes all**

Some of the filter operators in the preceding list are available for every field type. For example, **Exclude** and **Excludes all** only appear when filtering by multiselect field types.

Adding the field filters and using operators is not the full story. When using more than one filter, you can also set proper filter logic. It's very easy, so let's do this step by step:

I.   Create a report with at least two filters.

II.   Press the **Add Filter Logic** option placed in the **Filters** subtab, as shown in the following image:

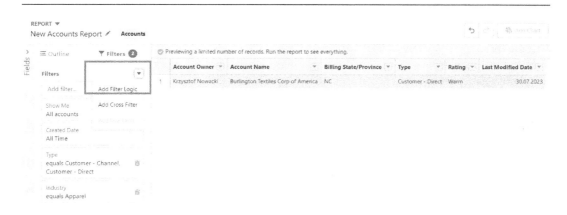

Figure 8.9: Adding report filter logic

3. Set the desired report filter logic. As the default, you will see that the **AND** is used, but you can replace it with other operators such as **OR** or **NOT**. Using those will influence the report result list:

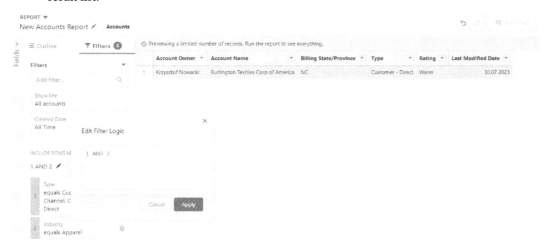

Figure 8.10: Report filter logic example

4. Click the **Apply** button to see the results.

5. Play around with **Filter Logic**; for example, change AND to OR to see how the record result list will change.

If you thought that's all there is regarding filters, then well you are wrong. There are some other features worth mentioning. One of them is called **Cross Filters**. Cross filters give you the option to create a report that will show records with or without relations to the records of another object. For example, you can create a report showing accounts without opportunities, as shown in the following image:

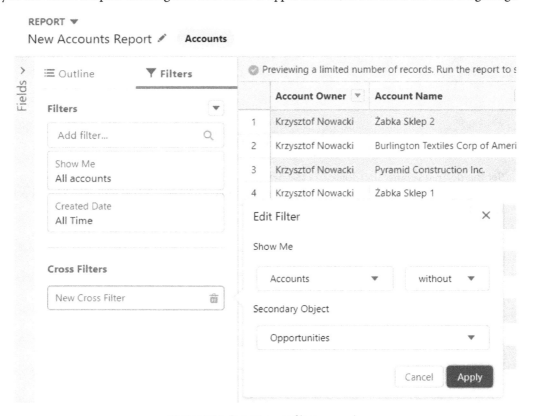

Figure 8.11: Report cross filter example

Although the cross filter is a somewhat hidden feature, I use it frequently, as many of my report-related requests involve displaying records that are either connected or not connected to other records. Thanks to cross filters, you can create this type of report logic. Let's now see how to use report groupings.

Report groupings and joining reports

Report Group Rows is a feature that helps you group the report data in any way that you want. You can pick any object field and group the report data according to this field. In the following example, I have grouped the accounts according to the **Account Owner** and **Account Type** fields. As you can see, the grouping structured the data and created separated groups, where you can see the data separated by **Account Owner** and **Account Type**. Additionally, this grouping gives you the possibility to create report charts. There is no way to create a report chart if the data are not grouped:

Figure 8.12: Report rows group

When grouping, we can use two types of grouping:

- **Group Rows**: You can use a max. of three levels
- **Group Columns**: It creates a matrix type of grouping

Report grouping is an easy but very useful and important function that will allow you to present your data in an organized way. Let's now see how Buckets, Summary, and Row-Level Formulas support this.

Report buckets, summary, and row-level formulas

By employing a bucketing method, you can swiftly classify report entries without the need to generate a formula or custom field. By establishing a bucket column, you delineate numerous categories (buckets) designed to cluster report data. To create a bucket column, click **Add Bucket Column**, which is hidden under the arrow in the report **Outline** section. Please see the **Bucket Column** example in the following image. As you can see, we have assigned the different opportunity stages into two separations: **Initial Stage** and **Advance Stage**. Some opportunities are assigned as **Initial**, whereas others are assigned as **Advance**. Please check the example in the following image to see how buckets can be set up:

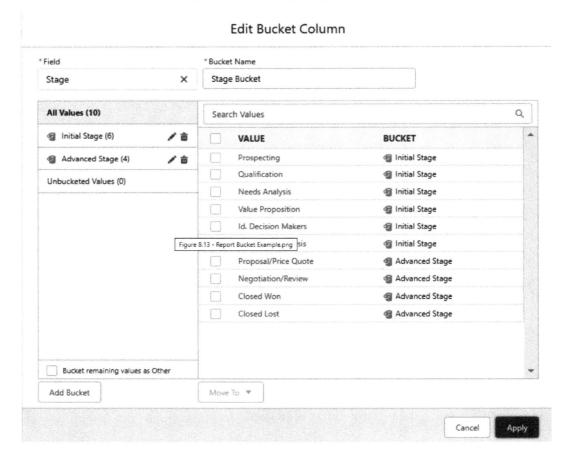

Figure 8.13: Report bucket example

Similar to any other column within your report, you retain the ability to arrange and sift through data based on these bucket columns. Please see the bucket grouping in the following example:

Stage ↑ ▼	Stage ▼	Type ▼
	Id. Decision Makers	Existing Customer - Upgrade
	Value Proposition	Existing Customer - Upgrade
Subtotal		
Advanced Stage (15)	Negotiation/Review	New Customer
	Closed Won	Existing Customer - Upgrade
	Closed Won	Existing Customer - Upgrade
	Closed Won	New Customer
	Closed Won	Existing Customer - Upgrade
	Closed Won	Existing Customer - Upgrade
	Closed Won	New Customer
	Negotiation/Review	Existing Customer - Upgrade

Figure 8.14: Report bucket grouping

When you know how to bucket the data, we will then see how to use features similar to formula fields, which are called **Summary** and **Row-Level** formulas.

Summary and row-level formulas

In addition to standard columns, which come directly from the field values stored for the Salesforce objects, you are able to create additional columns that are formula-based. Those columns are calculations made in the formula editor and help you add more valuable data to your reports. At the moment, you are able to create two types of formulas: **Summary** and **Row-Level**. Let's describe them in more detail in the following lines:

- **Summary**: This helps you to compute extra aggregate values derived from the numeric data present in your report. Next to the conventional summaries, you have the option to include up to five distinct summary formulas within summary and matrix reports. These formulas enable the creation of calculated summaries for your numerical fields. In the following example, you see the **Summary-Level Formula** for calculating the opportunity win ratio, displaying the score in numbers (using two decimal points) for the **Grand Total Only** (meaning the final summary of the report data):

Edit Summary-Level Formula Column

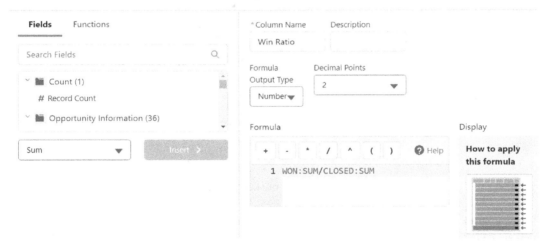

Figure 8.15: Report summary formula example

As you can see in the preceding image, you can use report summary formulas not on each row level but on specific row levels, such as All Summary Levels, Grant Total Only, and Selected Groups. Sounds like a limitation? Yes, it is. But don't you worry. Salesforce has got you covered and gives you the power to create a formula on each row when using the **Row-Level** formula. Let's learn more about this feature in the following text.

- **Row-Level** formula: Incorporating a row-level formula introduces a column dedicated to row-level formulas within your report, performing calculations on each individual row present in the report. Let's see an example in which I've calculated the time to close on each opportunity in my report:

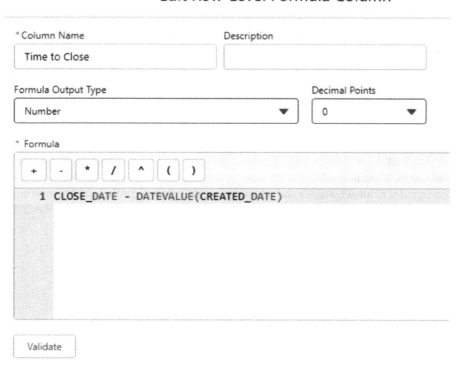

Figure 8.16: Report Row-Level formula example

As you can see in the image, we only subtracted the opportunity creation date from the planned closing date. In the meantime, we have also changed the Created Date value type to Date to be able to carry out subtractions, as this calculation requires the same types of data, and originally, Close Data was a data field, and Created Date is a Data/Time field. When you validate and apply your formula, you should be able to see the calculations in each opportunity report row in a similar way to that shown in the following image:

Stage ↑ ▼	Type ↑	Opportunity Name ▼	f_x Time to Close ↓ ▼
Closed Won (16)	Existing Customer - Upgrade (10)	United Oil Standby Generators	392
		University of AZ SLA	154

Figure 8.17: Report Row-Level formula as a column

As you can see, the **Summary** and **Row-Level** formulas represent fantastic add-ons to Salesforce reports. Thanks to them, you do not need to create additional formula fields for objects but can, instead, do some necessary calculations directly in the reports. When we have understood how to add more information to our reports, we will see how to select data from the crowd using **Conditional Formatting**.

Conditional Formatting

Sometimes, you might like to highlight some of the important numbers in your report. Does Salesforce give you this possibility? Yes, it does! For this reason, you can use the **Conditional Formatting** feature. **Conditional Formatting** is available in matrix-type reports or any other format where you have summarized number fields. Please see the following image, where **Conditional Formatting** for the opportunity **Sum of Age** field can be seen. You can create the data ranges and assign the color to it:

Figure 8.18: Report Conditional Formatting example

When you finish the Conditional Formatting setup, you will see that, on the report UI, the corresponding numbers will be marked with the appropriate colors, as seen in the following image:

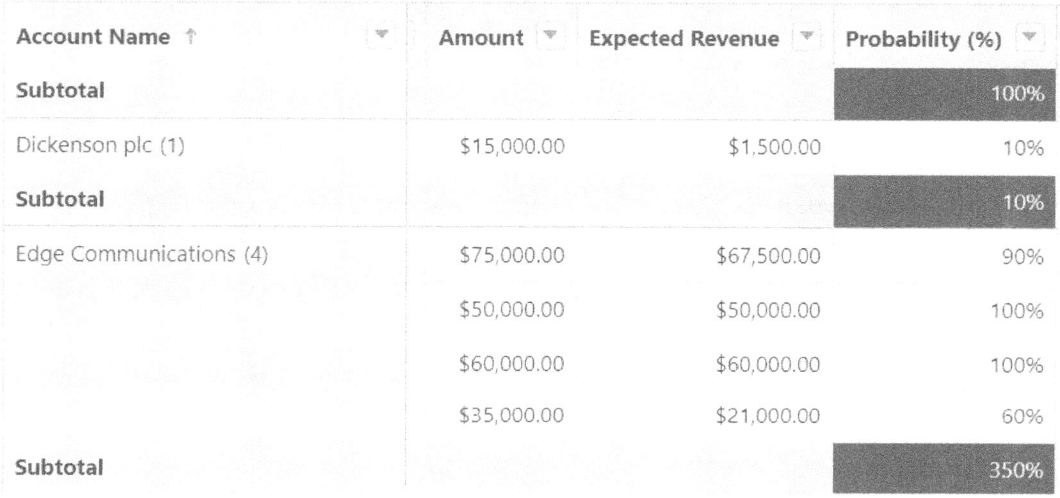

Account Name ↑	Amount	Expected Revenue	Probability (%)
Subtotal			100%
Dickenson plc (1)	$15,000.00	$1,500.00	10%
Subtotal			10%
Edge Communications (4)	$75,000.00	$67,500.00	90%
	$50,000.00	$50,000.00	100%
	$60,000.00	$60,000.00	100%
	$35,000.00	$21,000.00	60%
Subtotal			350%

Figure 8.19: Conditional Formatting on Run Report UI

Conditional Formatting can be useful to highlight detailed data, but sometimes, we want to omit the details and show only subtotal or totals. Let's see how to do it in the next paragraph.

Detail Rows, Subtotals, and Grand Total

Salesforce reports can become massive when showing a large amount of data. Sometimes, you want to report only the needed data, and it is not necessary to show all the data details to the user. For this reason, Salesforce reporting gives you the power to toggle the visibility of three report elements:

- **Detail Rows**
- **Subtotals**
- **Grand Total**

How do we use these features? Let me give you a simple example. Sometimes, you may want to hide **Detail Rows** because you only want to show **Subtotals** and **Grand Total** or even **Grant Total** only. You can switch off the needed details, and that's it! Play around with this feature and see the results for yourself, as the usage of it depends on the reporting needs.

Adding report charts

Report charts are a great addition to table report data. Illustrating information helps a person understand it more easily and see the most important information. To add a chart to a report, you need to create the grouped report. Please follow the following steps to add the report chart to the report:

1. Create a grouped report.

2. Click the chart icon on the right-hand side of the report.

3. Choose a chart type in the **Display As** section. You have the following options:

   - Bar

   - Column

   - Stacked Bar

   - Stacked Column

   - Line

   - Donut

   - Funnel

   - Scatter Point

Figure 8.20: Report chart setup

4. Add additional information, such as **Show Values**.

5. **Run** the report to see the outcome. It will look similar to the following image:

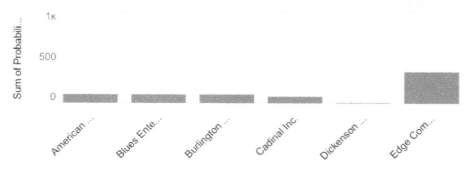

Figure 8.21: Report chart example

Choosing the best graph is half the battle of effective reporting. Salesforce allows you to choose several graphs. How you want to show the data depends on you and your business requirements. Now that know how to build the report the way you want, let's see how to export this data from Salesforce.

Exporting report data

We have already learned about exporting data via Salesforce reports in *Chapter 3*. To learn more details about this feature, we encourage you to return to this chapter. As a reminder, we will only mention here that you have two options when exporting data: **Formatted Report** and **Details Only**, and a few formats and encodings that you can use.

Report types

When creating Salesforce reports, you might notice that you are using some predefined report templates. You have also seen that in those templates, you have some predefined columns, available fields, and objects. Those templates are named report types, and besides the predefined ones, you are able to create your own custom report types in just a few clicks. The need for custom record types may appear when there is no standard record type available that connects some specific objects, or together and the fields are not available.

Some important limitations related to custom report types are the following:

- There is a limit of a maximum of four objects that can be associated with custom report types

- A custom report type can contain up to 60 object references

- You can add up to 1,000 fields to each custom report type

When creating custom report types, you can decide about object relationships. You can choose the option **Each "A" record must have at least one related "B" record**, which will serve you the report showing only records connected to each other via the object dependency. An example of this situation is shown in the following image:

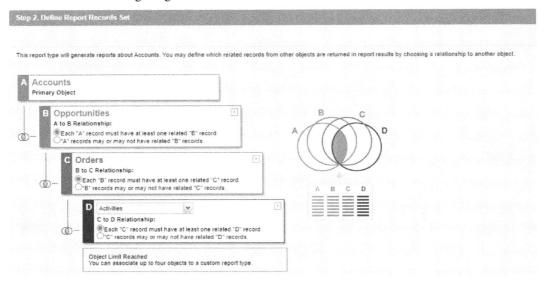

Figure 8.22: Custom report type "Must" relationship

You can also choose the second option named **"A" records may or may not have related "B" records.**, which means that the report will also show you those records not connected to each other:

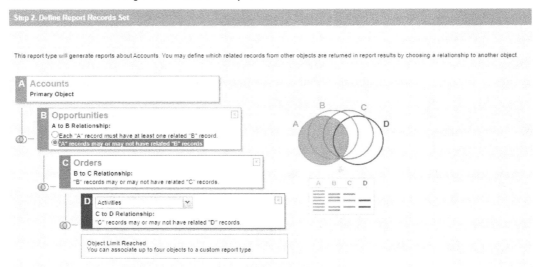

Figure 8.23: Custom report type "May" relationship

Of course, there can also be a mix of both options, so each time you use it, you can play around with the report type builder to obtain the needed solution. In addition to the object setup, you can also decide which fields will be available and how they will be available in your custom report type; just use **Edit Layout** as the final step in the custom report type setup. In addition to this mode, you also have the **Preview Layout** mode available, as shown in the following image:

Figure 8.24: Custom report type with Edit and Preview Layout options

Building custom report types is easy, but to do so definitively, you need to have strong knowledge of your Salesforce organization's data model. Without this knowledge, it may take you a while before you are able to create the desired report type.

> **Important note**
>
> If you want to create the Salesforce report type that will not be accessible to other users, you can set its deployment status to **In Development**. When choosing this status, the report type, along with its reports, remains hidden for all users, except for those users possessing the "Manage Custom Report Types" permission.

Of course, practice makes progress, so working with custom report types will also help you learn more about the logic of your data model. I encourage you to try building your own report types. There's a lot you can learn from doing so.

In the preceding subchapters, you have learned how to master Salesforce reports. You've discovered how to manage columns, apply filters, group the data, create report formulas, incorporate charts, efficiently export reports, and create custom report types. These skills empower you to analyze, visualize, and share data effectively, aiding informed decision-making within your organization.

Now that you know almost everything about Salesforce reports, let's learn how to build informative and useful Salesforce dashboards from this. Dashboards not only check your sales progress but also give live information and help a company grow and make decisions.

Dashboards

To even start discussing dashboards, you need to know what they are. Let's begin with the idea that dashboards are somewhat like a data collection center. For example, in movies, when they show the main villain's room, there is usually a huge wall with various indicators. Now we can have such a wall (and here is the moment for an evil laugh!). To illustrate this more, look at the following image:

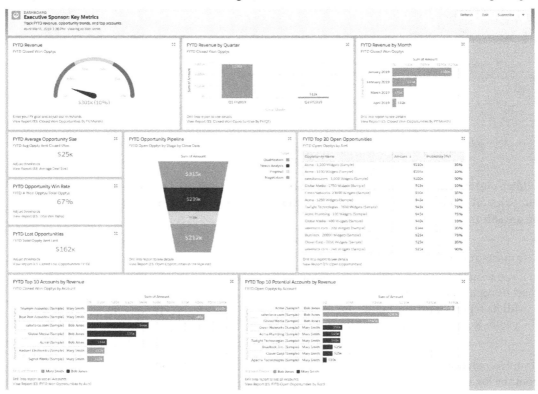

Figure 8.25: Executive dashboard (sample)

As you can see, there are a lot of bars, colors, and data. So, in simple terms, a dashboard is a tool that enables the presentation of data from various reports in one place. For users, they often serve as a main source of information, which is immediately available, goes through many objects at the same time, and provides knowledge on many levels of the business.

It is worth mentioning that companies usually have several dashboards, and there is a simple explanation for this. Large, medium, and small companies have people responsible for specific areas in the company, such as sales, marketing, finance, and management. Each of these departments needs completely different information. For the marketing department, sales charts are not necessary, and for finance, a summary of successful marketing campaigns is not needed. That is why system admins build dashboards filled with different data.

Now that you know what a dashboard is, let's move on to where to find them. It's amazingly simple; go to the **App Launcher**, the icon with nine dots, under which you can find all the applications. There, type Dashboards in the search bar, and voilà. A common practice is to implement dashboards on the homepage. In this way, right after logging into the system, users receive insights into all the necessary data. I remember working at a help desk, where I first learned about Salesforce. We used it as our main ticketing tool. Right after logging in, I had my own dashboard where I could see my unresolved cases, new issues assigned from users, my monthly performance, and many other crucial elements.

To create a new dashboard that meets user requirements, you need two elements: reports and what you want to extract from them. Let's assume we already have these reports, as you have already learned a lot about them in the previous part of this chapter. Creating a new dashboard is exceptionally simple; we go to **Dashboards** in the **App Launcher** and click **New Dashboard**. However, the fun part really starts now. Let's go through it together step by step:

1. After clicking on the **New Dashboard** button, you will receive a new popup with fields such as **Name**, **Description**, and **Folder**, as shown in the following image:

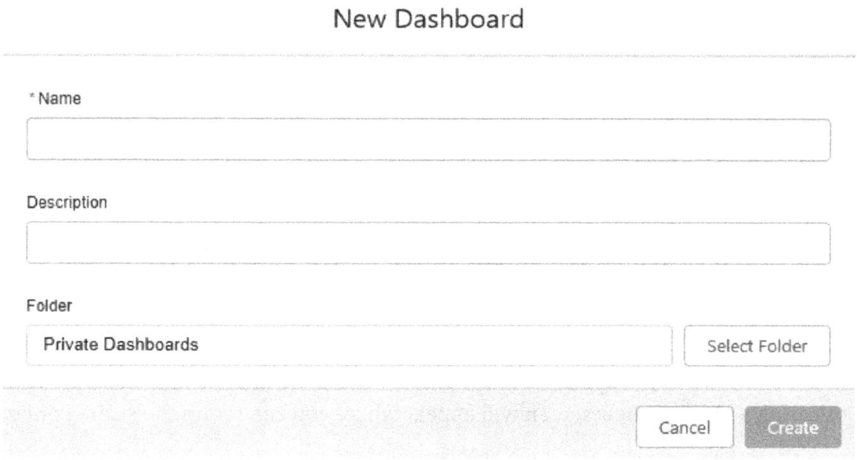

New Dashboard

* Name

Description

Folder

Private Dashboards   Select Folder

Cancel   Create

Figure 8.26: New Dashboard popup

Enter all the necessary details and choose a folder to save it in. The handling of folders and their sharing is identical to that of reports. After clicking **Create**, you will receive an empty board, which you will soon fill with charts.

Before creating new charts, let's first enter the settings. Click on the gear icon. A **Properties** popup will appear on the screen, where you will find the name, description, and folder. It also includes options such as **View Dashboard As**, which is the choice of perspective from which data on the dashboard will be displayed, and **Dashboard Grid Size**, which determines how many columns the dashboard grid will have. At the bottom, there is the choice of the dashboard's color palette, and you can get creative; look at the following image to see this:

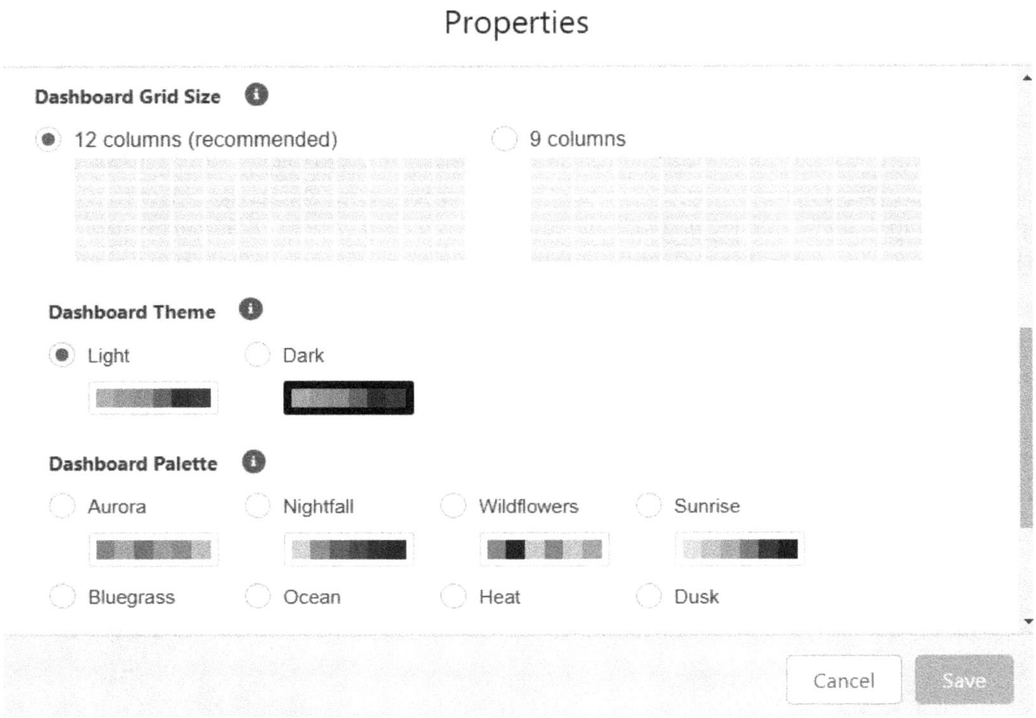

Figure 8.27: Dashboard settings

2. The next step is to finally create new charts. To do this, click on the + **Widget** button, and among the three options, choose **Chart or Table** (the other two options are **Text** and **Image**).

    Now, you can select a report. Remember, for a chart to show something meaningful, group certain values. Without grouping, you can only use a table, which is essentially like a list view. After selecting a report, a new screen will appear where you can set up the entire configuration.

At the top, your chosen report is displayed, along with other options:

   A.   **Types of charts**.

   B.   **Y-axis** and **X-axis**, which define values on the Y-axis and X-axis

   C.   Next to the Y-axis, you have the option to add two data groups for analysis, with the first being the main one

   D.   The **Display Units** field determines the units for displaying data

   E.   The **Show Values** checkbox decides whether values are displayed on the chart

   F.   **Show Chatter photos** defines whether the chart should display images from Chatter

   G.   **X-Axis Range** is the choice of the X-axis range; you can choose whether it is automatic or custom

   H.   **Decimal Places** determines the number of decimal places.

   I.   **Sort by/Then Sort By** is used to specify the data-sorting criteria

   J.   **Custom link** allows for the addition of a link to the chart.

   K.   **Max Groups Displayed** is the maximum number of data groups displayed on your chart.

   L.   **Title/Subtitle/Footer** is used for the customization of these elements

   M.   **Legend Position** is used to determine the position of the legend

   N.   **Component theme** is the choice of the component theme, which can be light or dark

3.   After choosing these data and accepting the visualization on the right side, click the **Add** button and choose where the new chart should be placed.

Once we fill the entire dashboard, click **Save** in the upper right corner, and you can enjoy your new dashboard.

As you can see, creating a new dashboard is not an exceptionally difficult task. Before starting work, it is important to focus on defining the goals and metrics of your new dashboard, without which you cannot start configuring charts. When creating a new chart, there is a **Display As** field, where, as I mentioned, you can find types of charts, as shown in the next image:

Figure 8.28: Types of charts

Let's start with the highlighted type:

- **Bar chart** is designed for comparing different groups of data.

- **Column chart** is very similar to a Bar Chart, but with a vertical layout.

- **Stacked Horizontal Bar Chart/Vertical Bar Chart** is for those who love stacking data in one chart. It is useful for analyzing cumulative data.

- **Line Chart** is excellent for showing trends and changes over time.

- **Donut Chart** is used to represent percentage shares.

- **Metric Chart** is a type of chart used to present single key values.

- **Gauge Chart** looks a bit like a speedometer but is used to represent progress relative to a set goal.

- **Funnel Chart** is used to show sales processes and conversions.

- **Scatter Chart** is used to show the correlation between two variables.

- **Lightning Table** is used to present data in a table format.

 Administrators have many options for presenting data, but is that everything after creating such a dashboard for users? Can we close the door behind us and take the rest of the year off? I am afraid not because a few unexpected situations may occur. If your user contacts you saying that one of the charts is not working, you need to know that something is happening. What to do then? It is best to use the **Login as a User** option, which we discussed in earlier chapters, and see what the user sees. You may not always see what they "see." The reason for an error in displaying a chart could be a lack of access to data resources or the removal of a grouping in the report. Verify the previous state of the report and see what has changed. Before making any configuration changes, consult the granted/ revoked accesses, as the change might have been intentional.

Best practices in designing dashboards

Remember, cleanliness and simplicity allow users to extract the most important data. Do not remove legends from charts; they help users understand what they are looking at. Tailor the dashboard to the user's needs and knowledge level, and if they want simpler charts, give them that. Proper visualization is extremely important; choose chart types that best fit the data.

Remember when I mentioned there could be several dashboards for different teams? But how is this done so that one application has two different home pages, each assigned to a specific profile? Very easily. You use the Lightning **App Builder** for this. To do this, go to the homepage, click on the gear icon, and enter **Edit Page**. Once you have created several types of dashboards, select the **Dashboard** component, and place it on the homepage. Then, choose the dashboard to be displayed. After selecting the appropriate dashboard, save the page with the **Save** button, and then click on **Activation**. In the new window, select the **App** and **Profile** options, where you choose the application and profile that the dashboard will display. Repeat the process for other profiles that require dedicated dashboards.

I hope working with this solution will be enjoyable for you. Personally, I think Salesforce did an excellent job of designing dashboards; the design and configuration are exceptionally intuitive. I am sure that after a few set dashboards and further training, you will become a master of dashboards. So, as Mr. Miyagi from Karate Kid said, "*Wax on, right hand. Wax off, left hand. Wax on, wax off. Breathe in through nose, out the mouth,*" and now get started on making new charts.

Summary

As you know, reporting combined with smart data visualization is one of the most essential skills today, as "without data, you're just another person with an opinion."

In this chapter, you read about reports and dashboards, two elements masterfully designed by Salesforce. From what you have read, you now know how to create a new report, put it in a folder where no one will see it or share it with many people, create appropriate filters and groupings, and add new fields. After creating the report, all that remains is to create a chart for it, where we extract the most important data for the user and visualize it in an accessible way for them. I wish you the obtainment of a black belt in creating reports/dashboards. After going through this chapter, you're now ready to use these skills to report your Salesforce data better and make smarter decisions at work.

In the next chapter, we will open the doors to the ready-made Salesforce solutions for you to use. You will learn what **AppExchange** is, how to install an application from it, discover interesting functionalities from Salesforce Labs, and much more. So, see you in the next chapter!

9

AppExchange
and Custom Applications

In the last chapter, you learned about reports and dashboards, an incredibly important part of Salesforce, loved by many. In this chapter, we will focus on existing solutions, but these won't be the default solutions that come with Salesforce Out-of-the-box. Instead, we will explore dedicated solutions for specific functionalities. In this chapter, you'll get to know AppExchange, what it is and why it's important, what you can find there, and how to use it.

The chapter will be divided into three sections:

- Introduction to AppExchange
- Salesforce Labs Apps
- Installing and Managing App

Introduction to AppExchange

Do you like shopping? I think everyone has their favourite store – some love to go Saturday shopping in a clothing store, others prefer a bookstore, and yet others opt for a computer store. Just like with tastes, scents, or favourite colors, we have our favourite places to shop. But have you ever considered a store to be a friend? I do not think so. However, with AppExchange, you will definitely hit it off. AppExchange is a special place in Salesforce where you can find custom solutions. On your phone, you have either the App Store or the Play Store, each filled with numerous apps designed to make your life easier or more enjoyable. Similarly, AppExchange is a digital marketplace for Salesforce applications. This platform allows you to search for, install, and review applications. You will find apps that expand the functionality of your system without the need for reconfiguration or development.

Where to find AppExchange? In our beloved **Setup** – yes, that is exactly where it is located. Type AppExchange in the search bar and click on the search result. A new tab with a library of solutions will open. Or, more simply, click on or enter the URL `https://appexchange.salesforce.com`.

What can you find there? I think the answer "everything" will not be misleading. But let us start from the top of the page, which looks exactly like the one in the next picture.

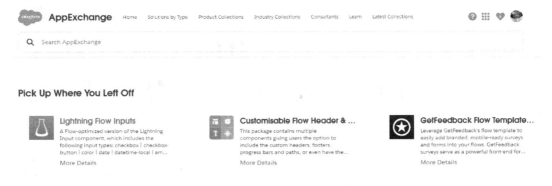

Figure 9.1: AppExchange home page

As you can see, at the top, you have the option to select from the necessary categories, which we will list below:

- **Solutions** by type is a place where you can find:

  - **Apps**, where you will find pre-integrated solutions for most SF clouds.

  - **Bolt Solution**, which are solutions based on Experience Cloud, allowing you to quickly launch a portal/page integrated with your Salesforce database. Examples: Alumni Engagement Pack for handling alumni, PatientX Telehealth Solution for engaging with patients, or Supplier Relationship Management.

  - **Flow Solutions**, which are ready-made flows waiting for your installation.

  - **Lightning Data**, a collection of data that can enrich current records in the system, introduce proper verification, and provide additional data from external sources.

  - **Components**, which are components you can use while building page layouts in Lightning App Builder.

- **Product collections**, which are apps sorted under appropriate clouds, ensuring you do not find apps dedicated to Service Cloud in the Sales Cloud section.

- **Industry collections**, similar to the above but in this case, the apps are grouped into specific business sectors.

- **Consultants**, where you can find a Salesforce partner. If your company is implementing SF, it is listed here.

- **Learn**, a collection of knowledge articles for you to read.

- **Latest collection**, which are collections of software on a specific theme. In this case, Salesforce itself creates these collections and makes them available to users.

Of course, these are certain sets of software that you can use. However, as often happens, you might be looking for something very specific in such stores. You know what you want and why you came here. What then? (I know, silly question because of course, you know what to do) You click on the search bar and enter the product you want to find. Salesforce did not disappoint us as it also introduced several filters that will help us find what we are looking for.

The first two are those that are highlighted in the next image.

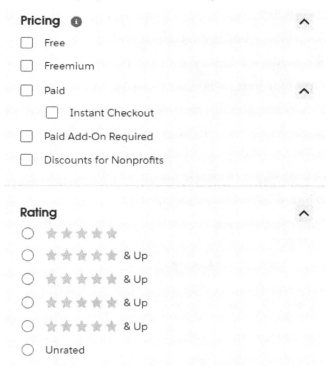

Figure 9.2: AppExchange Filters

In the Pricing filter, we can specify several types of application fees:

- **Free**, meaning there is no charge.

- **Freemium**, a type of application where you get some functionalities for free. However, to use the full functionality, you must pay.

- **Paid**, which are those applications that require payment. Additionally, paid has a sub-option called Instant Checkout for applications where you can pay with a credit card or bank transfer.

- **Paid Add-On Required** for applications that require payment for an external product, one that is outside of SF.

- **Discount for Nonprofits**, which are applications that offer discounts for NGO organizations.

Below the pricing, you will find ratings, which I do not need to explain. The standard star system works here just like everywhere else; the more stars, the better. And those with the maximum number of stars are the ones you should ideally be looking for.

Moving down, the next filters include:

- **Business Need** for products dedicated to business needs.

- **Solution Type** for the type of solution you are looking for, such as Flow, Component, etc.

- **Product** for the type of SF product like Data Cloud, Experience Cloud, and many others.

- **Salesforce Labs**, a cute little checkbox that will search for products produced by this team.

- **Supported Features** for applications that support specific elements in Salesforce, e.g., Person Accounts or Multiple Currencies.

- **Salesforce Edition**, which you can find in Company Information.

- **Industry** for products dedicated to specific industries.

- **Job Type** if you are looking for a specific role from a Salesforce partner, such as Administrator, Consultant, or Developer.

- **Language** for products in a specific language.

- **Impact** for products whose providers align with certain values, such as the 1% pledge program where members donate 1% of capital, revenue to community purposes.

In each application from AppExchange, you will find several sections, as shown in the following image:

Figure 9.3: AppExchange App Sections

I would like you to know what to look for in such a description and how to carefully check if an application is a desirable choice. Always check the rating, as it will tell you what users really think about the application. Then, verify if the product is compatible with the cloud you want to use it for and if it has any specific requirements. Of course, it is worth reading the description, familiarizing yourself with the screenshots, and watching the product demo, which you can find on the main page under the Overview tab. If you scroll down, you can read the Highlights, where the producer lists the most key features of their product. Ah yes, money, let us talk about money (Scrooge McDuck would be quite irritated by such extravagance); always check the product's price. You learned about filters above; if you want to be 100% sure that the product is free, then apply it. However, you will always find this information on the main page; it is quite visible. Then you can move on to the Reviews tab. Thanks to this, you can sift through the marketing language of the creators and get real user opinions of the application. The last tab, More Details, includes equally crucial elements. In the Salesforce Edition section, you will know which editions the application fits. Under App Details, you will find the date of the first and last release of the app. Below that are Package Contents and Lightning Components, where you will find out what will be added to your system after installing the product - whether it be custom objects, fields, applications, or what components.

And the last piece of our puzzle is what we all love to do before any movement with an external application - reading the documentation. Yes, right there, at the very bottom, you will usually find links to the documentation. Happy reading.

But is everything as good as it looks, or is it really that good? AppExchange is a brilliant move by Salesforce. With so many available solutions, you do not have to build every solution from scratch; you can often save a lot of time thanks to the available products. So, before your developer starts developing a new product for your system, sit down with your Business Analyst, review what the client wants to achieve, and look for an available solution. Even if the application is paid, these are often relatively small amounts of money compared to the time of our developer, architect, and many others who can build a custom solution from scratch.

> **TIP**
>
> Frequently, applications like Mailchimp, Eventbrite, ActionKit, and many others have a connection with Salesforce. Just search your favourite search engine for phrases like Application Name Salesforce Integration. This way, the search engine should immediately suggest whether such integration is available or not. If you get a positive result, immediately check in AppExchange if such a connector exists, allowing you to smoothly connect the two solutions.

In summary, with AppExchange, just as you search for all the most important applications for your new phone, spend a few hours here, look around this virtual store, and maybe you will find something that will streamline your work. And now let me introduce you to one of the coolest teams, in my opinion, producing applications, components, and many other things in AppExchange. Meet Salesforce Labs.

Salesforce Labs Apps

Let us start with what Salesforce Labs is. It is not something like the Umbrella Corporation from Resident Evil, and that is for the best - a zombie apocalypse would be closer than we might expect. Salesforce Labs is a group of exceptional people at Salesforce who design and create solutions available on AppExchange. These solutions are often provided for free to all users and bring a lot of innovation to our systems. Personally, I am a fan of their solutions and the fact that they provide them entirely for $0. Salesforce Labs, as the team could also be called, gives Salesforce employees the opportunity to develop their own ideas and turn them into reality through their initiative.

With their applications, they demonstrate the platform's capabilities and its flexibility in adapting to the increasing business needs worldwide. Salesforce Labs engages in developing its user community, promoting innovation and a creative approach to the system.

I would like to show you the TOP 5 solutions released by Salesforce Labs. This TOP 5 is, of course, a subjective assessment, and you will surely find your favourite applications by checking all the available solutions.

- **Salesforce Adoption Dashboards.** This solution is recommended for anyone who wants to see how users have adapted to the new environment. For example, if your company is transitioning from Excel spreadsheets to Salesforce, do not hesitate. Try out the Salesforce Adoption Dashboard, which will give you insights into information such as login history, KPI indicators, the usage of individual Salesforce elements, and even user activity itself. This tool is not just for Administrators; higher-level individuals who want to see the return on investment in Salesforce can also use it. With this dashboard, you will know which user claiming they cannot log in was not lying. In the next section, you will learn how to install the Salesforce Adoption Dashboard. It looks exactly as shown in the following picture.

Figure 9.4: Salesforce Adoption Dashboard overview

- **Query Studio for Marketing Cloud** is a tool exclusively dedicated to Marketing Cloud. So, if you have this cloud in your organization, talk to its analysts and marketers; they will surely be grateful. This solution allows for quick and flexible execution of complex queries in Marketing Cloud, quite similar to SQL Server Studio or MySQL Workbench. Users can use this tool to create and test SQL queries directly in Marketing Cloud, enabling rapid data analysis. Look for the application looking like the one shown in the next picture.

Figure 9.5: Query Studio for Marketing Cloud

- **Time Wrap** is an application that allows you to visualize related records on an interactive timeline. This application provides a comprehensive view of records such as Accounts, Contacts, Leads, and Cases. Users can scroll through the timeline and click to get a full picture of their customer data. It is particularly useful for understanding complex relationships and interactions with customers, making it easier to comprehend the history and activities associated with a client. The application looks like the one shown in the next picture.

Figure 9.6: Time Wrap overview

- **Agile Accelerator** is an application that enables you to manage projects in accordance with Agile methodology. This tool allows for the management of backlogs, user stories, sprints, and other elements of Agile project methodologies. Agile Accelerator facilitates collaboration among entire teams on projects. It organizes work and, most importantly, makes it efficient, partly due to the ability to facilitate communication within teams. Agile Accelerator allows you to plan sprints, track project progress, and manage tasks. The Agile Accelerator looks like the one shown in the next picture.

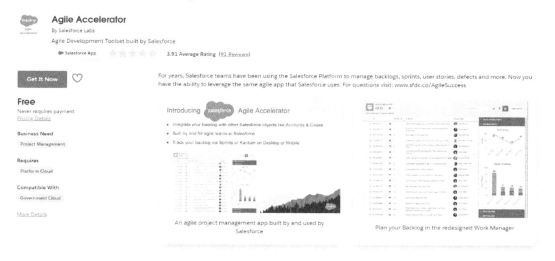

Figure 9.7: Agile Accelerator

- **Data Quality Analysis Dashboards**, last but not least, is an application that allows for in-depth analysis of your system's data and improvement of their quality. With a set of dashboards that analyze data quality in records, you can identify missing or incomplete data. How often have you encountered data entries labeled with 'QWERTY' or 'Lady Jane Doe'? This application enables quick identification and resolution of data issues. While it may not bring peace and harmony, it certainly ensures valuable data accuracy. This application is particularly useful in large organizations where a lot of data comes in and requires continuous attention. Look for a logo with a laboratory bottle and an application that looks like the one shown in the next picture.

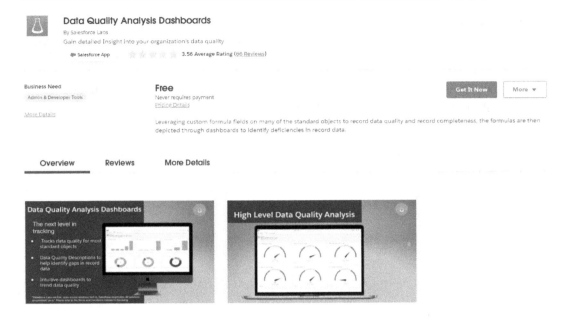

Figure 9.8: Data Quality Analysis Dashboard overview

I hope I have inspired you to explore the applications from Salesforce Labs. I'm confident that they can help you significantly improve your system. Such solutions save you a lot of time and nerves, and the result is amazing. So, now that you know what AppExchange and Salesforce Labs are, the next step is to learn the secret knowledge of installing solutions available on this platform.

Installing and Managing App

In the previous sections of this chapter, we learn about the Salesforce AppExchange apps, and we get to know few examples of important and useful apps. Now let's learn how to install one of them. You will see how easy it is.

To learn how to install applications from Salesforce AppExchange, we will use a practical example of installing an application that may be useful to each of you. This application is called *Salesforce Adoption Dashboards*. We already learned about this application more extensively in the previous subsection, so let's briefly recall that the Salesforce Adoption Dashboards offer a view into pertinent user login history and trends, the embrace of crucial features like accounts and opportunities, and enhancements that significantly improve sales and marketing productivity. When you become a Salesforce Administrator this kind of Dashboard will become a must have sooner or later.

Alright, enough theory. Let's move on to practice and install the "Salesforce Adoption Dashboards." For the purpose of this presentation, I will be installing the mentioned application in the Salesforce Trailhead test environment. However, you can decide where you want to install it yourself. The installation steps will be essentially the same, except for the step where you choose the Salesforce environment to log in to install the application. Follow me step by step to learn how to install the *Salesforce Adoption Dashboards* application:

1. Find the "Salesforce Adoption Dashboards" application on AppExchange. Simply visit the website `https://appexchange.salesforce.com/` and use the search function to find the *Salesforce Adoption Dashboards* application. The app should appear in the search results as shown in the next picture.

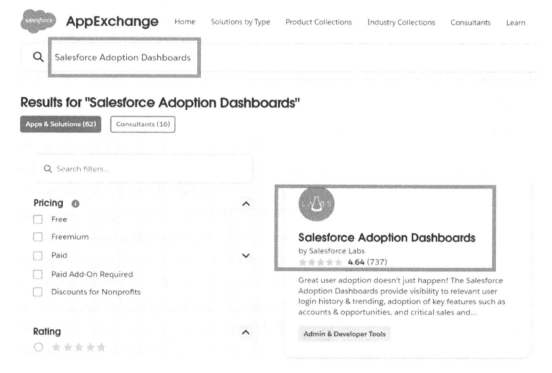

Figure 9.9: Searching for an AppExchange App

2. Click on the tile describing the **Salesforce Adoption Dashboards** application to view a detailed description and proceed with the installation.

3. Click the **Get it Now** button to start the installation process as shown in the following image.

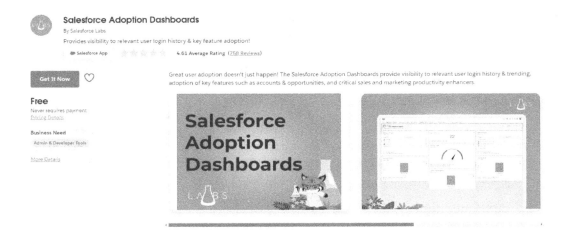

Figure 9.10: App Details and Installation

4. Choose where you want to install the app. As shown on the following picture you can choose between production or sandbox environment. In my example I will be choosing **Production** and will be installing the app in Salesforce Trailheads test org.

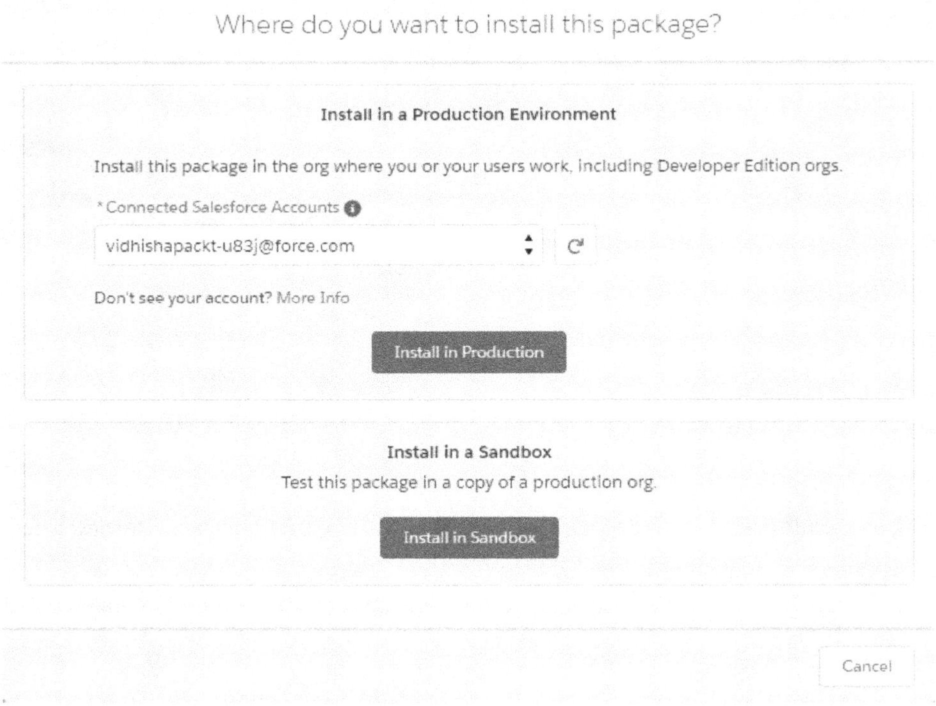

Where do you want to install this package?

Install in a Production Environment

Install this package in the org where you or your users work, including Developer Edition orgs.

* Connected Salesforce Accounts ⓘ

vidhishapackt-u83j@force.com

Don't see your account? More Info

Install in Production

Install in a Sandbox

Test this package in a copy of a production org.

Install in Sandbox

Cancel

Figure 9.11: App Installation Environment Options

5. Read terms and conditions and agree with them marking the checkbox displayed on the screen. Now you will be able to press the **Confirm and Instal** button. You will be redirected to Salesforce login page where you need to log-in to the environment where you want to install the "Salesforce Adoption Dashboards" app.

6. After successful login you will be prompted with **Installation Screen** where you can decide who will be able to see/use the app. Please choose **Install for All users**. Keep in mind that when installing a different app, you might like to install it only for Salesforce Admins or some other specific users but not for all of them so choose wisely. Click **Install** button to finalize the installation

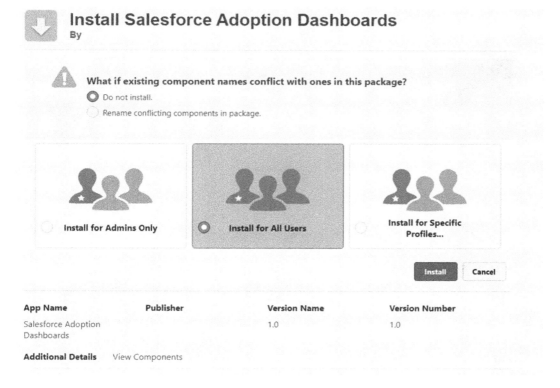

Figure 9.12: App Installation Screen

7. You will be prompted with app installation process screen. If app do not have many components inside it will be installed quite fast and you will be prompted with a success screen but if the app is more complex you will be informed that ot will take Salesforce more time to install it and you will be informed about the installation success in the dedicated email message. The app is an app which consist from Salesforce Dashboard/s, reports and just few new fields so it's rather fast to install so this time you should received the success screen after around one minute. As you can see on the following screen Salesforce will inform you that your instalation is completed.

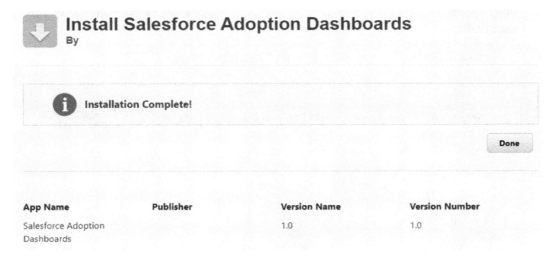

Figure 9.13: App Installation Completion

8. Go to the Salesforce to find and check the installed app. After clicking **Done** on the installation success screen you will be redirected to the Salesforce setup **Installed Packages** page where you can see all the installed app including ""Salesforce Adoption Dashboards". You can there check the detailed information about installed components or uninstall your apps.

9. Use installed app. Just go to the Salesforce Dashboards tab to check the installed Salesforce Dashboards. You should see three brand new Dashboards on the list. Each dashboards shows different data related to different Salesforce platform adoption related information:

    - **User Adoption (Logins)**

    - **Key Feature Adoption**

    - **Sales & Marketing Adoption**

10. Visit each Salesforce Dashboards or Analytics tab to check the installed Dashbaords data, discover what reports stands behind it and decide about needed adjustments. On the following screen I'm showing one of the installed Salesforce Dashboards which is **User Adoption (Logins)** Dashboard.

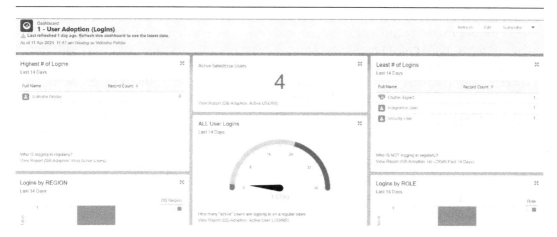

Figure 9.14: User Adoption Dashboard

This installed Sales Adoption Dashboards are off course a proposal of ready solution but can be off course upgraded with additional custom charts so finally you are able to develop your own personalized "Salesforce Adoption Control Center". I strongly encourage you to do so as the Salesforce adoption is one from most important things related with successful Salesforce implementation and platform management.

The AppExchange installation app example which we have just did was really "fast and furious" as it was quick, to the point and basically plug-and-play. Sometimes, however, the installation of an AppExchange app is associated with additional steps that need to be taken to make it available to users. Always read the instructions included with the application you are installing, as sometimes additional steps, such as creating additional Salesforce Permission Sets or adding them to users, may be necessary to use the application. In the case of paid applications, it may also be necessary to assign an application license to the user. Sometimes installation also involves adding additional Action Buttons or custom Salesforce Lightning Components on the user interface. All of this may sound complicated, but don't worry, as it should be thoroughly described by the company providing the application on AppExchange.

TIP

Sometimes the link to instructions related to the installed application is directly placed on the application's page on AppExchange. However, it occasionally happens that one needs to search more diligently to find such instructions and visit the application producer's website. I have also encountered situations where I contacted the publisher directly by sending them an email regarding this matter. Nevertheless, reputable application developers should provide instructions for their applications directly on Salesforce AppExchange.

Thanks to the AppExchange application ecosystem, Salesforce expands its capabilities in practically an infinite manner. You'll find on AppExchange applications related to human resource management, IT tasks, projects, contract management, and more. Sometimes, these are simple free solutions that are essentially plug-and-play, as their installation and usage are straightforward. However, on AppExchange, you may also encounter highly sophisticated applications that are additionally paid and may require several months of implementation. As you can see, you'll find various solutions there, but they all serve one purpose: expanding your environment more quickly so that you don't have to create everything from scratch. It's easier to install a ready-made application that has been tested and rated by other users than to reinvent the wheel yourself. I recommend keeping this last sentence in mind, as questions related to Salesforce AppExchange often appear in the Salesforce Administrator certification exam. The context usually revolves around the idea that it's better to check whether the desired functionality already exists on AppExchange before starting to create it from scratch.

Summary

Here we are, at the end of Chapter 9. Your knowledge of Salesforce is already quite substantial, and yet, this is not the end of the book. In this chapter, you learned that while Salesforce is indeed the best CRM in the world, it still doesn't have all the bells and whistles that a user might dream of. To address various needs of users, Salesforce supports itself with applications from the AppExchange store, the app market which belongs to the Salesforce itself, but you can find their apps that are created by Salesforce technology partners. Those partners are called by Salesforce ISV which stands for **Independent Software Vendor.**

After you found what the Salesforce AppExchange is, what kinds of applications you can find there, and what criteria to consider when searching for them we described examples of a special type of application available on AppExchange under the Salesforce Labs Apps label—applications created for Salesforce by Salesforce's own development teams. We reviewed the most popular and useful applications from Salesforce Labs, and ultimately, you learned how to install such an application on your chosen Salesforce org.

The Salesforce AppExchange marketplace continually evolves with new and innovative solutions, providing users with the flexibility to customize and optimize their Salesforce experience. The wealth of applications available on AppExchange allows you to tailor Salesforce to your organization's unique requirements, unlocking new possibilities and efficiencies extending Salesforce standard capabilities. Please keep this in mind when reading upcoming chapters of this book, where we will revisit the out-of-the-box Salesforce features related to Sales and Service Cloud. Now you know that even if something is not available in Salesforce out-of-the-box, it may be possible to add it by installing applications from AppExchange.

10
Service Cloud

In the previous chapter, you read about custom applications available in AppExchange and solutions created by a group of fantastic people – Salesforce Labs. But what if we need something more comprehensive that will allow us to launch our own help desk? That's what this chapter is about, namely, Service Cloud.

We will cover essential issues including the structure, application, and most important features. These elements will be discussed in the following sections:

- Introduction to Service Cloud
- Case management

Introduction to Service Cloud

Surely, at least once in your life, you have contacted a help desk. Remember that annoying hold music while waiting for someone to pick up? I do! I used to be on the other side of the phone, taking calls from clients (and I had put them on hold if the person on the other side was unpleasant. It taught me to be nice to help desk personnel). Interestingly, after picking up the phone, I logged the case in Service Cloud.

As mentioned in previous chapters, Salesforce has designed many types of clouds for different purposes, including Service Cloud, which makes life easier for managers and consultants working on the help desk.

Why did Salesforce decide on this type of solution? I think that question might be answered by the founder of Salesforce himself, but just like the ingredients of KFC's famous chicken seasoning, we will never know. There are only guesses. Business consists of two elements that are in complete opposition to each other – sales and complaints. So, if something has been sold in Sales Cloud and we can enjoy the income, let's make sure that in case of complaints, we manage it just as perfectly.

Now that we know (supposedly) why Service Cloud was created, let us focus on what Service Cloud really is. Let us uncover this secret knowledge. Service Cloud is a tool offering a range of solutions that help us achieve excellent communication with the customer. These tools make our customer service more personalized and effective.

To illustrate this even better, let us use the example of a newly formed company that will reserve a parking spot for you in the morning and release it when you come to pick your car up, called "*That's My Spot*" (I can already hear Sheldon Cooper from The Big Bang Theory saying it).

"*That's My Spot*" faces the following customer service issues:

- Managing different case issues that come from various channels (email, phone, chat, and social media)
- Staff do not have work aids such as automation or AI
- Agents must search for answers to questions themselves
- Every customer is treated the same – there is no distinction between corporate and individual clients

Luckily, someone recommended Service Cloud to the company. The company came to a meeting with a Salesforce partner and started talking. The team explained their problems, and right after that, they saw a sparkle in the consultant's eye; they knew he had taken up the challenge.

The consultant proposed the following:

- Thanks to Service Cloud, "*That's My Spot*" will be able to manage communication from various channels in one place. Whether it is a phone call from a customer, an email, a chat, or maybe a message sent through WhatsApp (as in the following image), agents will be able to receive them all on their SF account.

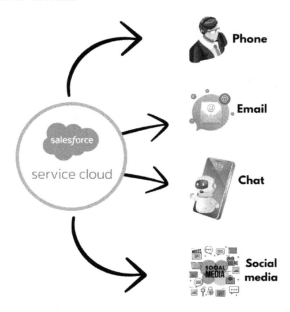

Figure 10.1: Service Cloud channels

- Thanks to Einstein's suggestions, an agent can respond to a customer's question more quickly. Additionally, any outdated case is flagged via automation as a high priority and reported to the manager of all the agents.

- A knowledge base available to everyone helps solve common problems, significantly easing the agents' work.

- Each customer, after verification, is assigned to the appropriate problem-solving path. Corporate customers have their own path, and individual customers have theirs.

So, as you can see, this is a cloud that helps those who help us. In summary, every company, just like "*That's My Spot*," wants to serve its customers at the highest level. If they could, they would hire a fortune teller with a crystal ball who could predict problems in advance (but no one has found one yet). Thanks to the many great solutions in Service Cloud, recruiters can breathe a sigh of relief and stop looking for someone with a crystal ball because they have a digital cloud that not only helps them improve customer service quality but also ensures customer satisfaction, loyalty, and company success.

And what does the architecture of Service Cloud look like? Surely such an efficient tool cannot have a simple architecture, and here we are right. Of course, it is not a maze without an exit; let me guide you through it.

First, Salesforce has designed each of its clouds in such a way that they can be scalable and expand freely in the future. So, if someone tells you that you have reached the end of configuration and expansion with Salesforce, you can confidently say that you have not even reached half of its potential. Service Cloud, as one of the main Salesforce clouds, is fully connected with Sales Cloud (which we will discuss in the next chapter) and other functionalities. This allows the agent managing a customer case to check their past purchases, resolved cases, and any previous communication had with them. Thanks to this coherence, the user has a comprehensive view of the customer in one place.

When talking about the structure of Service Cloud, I would like you to become familiar with some of the most important objects and connections between them. Let us start with the basic structure, which you can find in the following diagram:

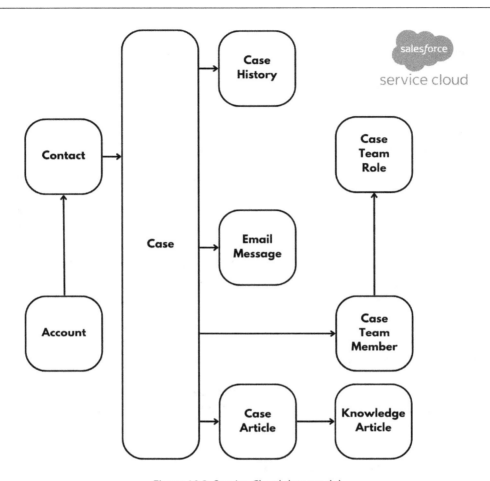

Figure 10.2: Service Cloud data model

I think we can skip the first two objects on the left because you are already a specialist in Accounts and Contacts, right? Cases are the fundamental object of Service Cloud. This is where you find all the details of your issues with a subscription, hairdryer, or a parking spot taken by someone else (we report this to "*That's My Spot!*"). Cases can come from various channels; we can distinguish between phone, email, chat, and social media. Cases are the central point for gathering information about customer problems. Around Cases, there are numerous brilliant functionalities, which we will examine in the following pages of this chapter. The following diagram is a breakdown of what we see in Cases by default.

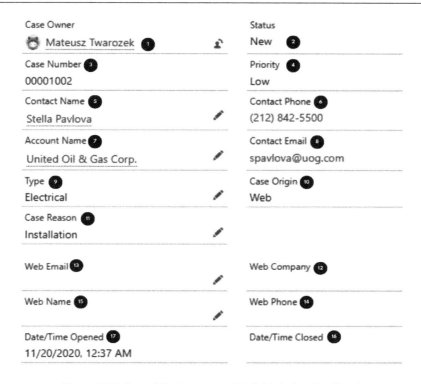

Figure 10.3: Out-of-the-box case with fields in Service Cloud

So, let us go through what is found in Cases:

1. **The Case Owner** is the person currently handling the case. Remember, at any time, the agent can be changed to someone else. In the case of a change of owner, there is also the possibility of changing ownership to a queue, which usually contains several users. Thanks to queues, there is no need to assign a case to one user. Queues facilitate the organization of group work and workflow management, ensuring simply that no task is overlooked. More about queues later in this chapter.

2. **Status** is a picklist with all statuses for a specific case. Remember that you can have several statuses for a case – these are called support processes. Using values from the picklist, you can easily match values to specific cases, as shown in the following screenshot. After selecting, you add them to record types.

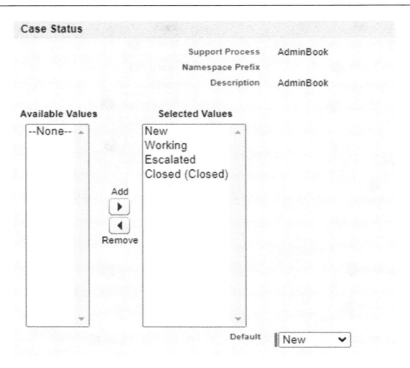

Figure 10.4: Support process

3. **Case Number** is usually a number automatically assigned to the case. These numbers serve as the name for this type of record.

4. **Priority** is simply the case's priority. Out of the box in Service Cloud, we have values such as **Low**, **Medium**, and **High**.

5. **Contact Name** is a lookup field that directs us to Contact object records. It lets agents know who they are conversing with.

6. **Contact Phone** stores the contact's phone number and is automatically retrieved.

7. **Account Name** is another lookup field, but this time it guides the agent to the Account object records.

8. **Contact Email**, similar to **Contact Phone**, stores the value provided in Email Contact.

9. **Type** is a simple picklist where the administrator can set case types. This allows an agent to select the appropriate type, which can later be used in case assignment or analytics.

10. **Case Origin** stores information about where the contacting person came from, whether it was a web form, phone, or email. These values can be filled in.

11. **Case reason** here you can capture why client contacted you. It's a picklist, you can change the values with a few clicks.

12. **Web Company**, 13. **Web Email**, 14. **Web Phone**, and 15. **Web Name** are fields that store information about the website of the contacting person. These fields are created in the out-of-the-box version.

16. **Date/Time Closed** is an extremely important field holding data on when the case was closed by the Agent.

17. **Date/Time Opened** is the field where the date of case creation is recorded.

Can Cases contain custom fields? Of course – they often even need to. Every company is a bit like a fingerprint: unique and unrepeatable. That is why they need different data on their cases. Creating new fields on a case is extremely simple, and we covered it in *Chapter 3*. The same applies to cases. So, if you are an admin in a footwear company, you can freely add fields for shoe size, a picklist for choosing material, and a multi picklist for selecting the issue/issues the customer is contacting customer service about.

Among the objects that play important roles in Service Cloud, I would like to highlight two additional ones.

One of them is **Work Orders**. This object is designed for service technicians or other specialists whose work is based on field operations such as repairs or maintenance. Work Orders allow for detailed planning of outing schedules, assigning human resources to them, and tracking their completion. This object is extremely useful for companies with field workers, making managing such tasks ridiculously easy.

Another incredibly crucial element of Service Cloud is the Knowledge Base, specifically **Knowledge Articles**. I remember when I worked at a help desk. Starting work in a company always involved onboarding, during which I was taught how to solve the most common issues. But seriously, who remembers all that? Amid the excitement of a new job and the fear of the unknown, someone tells you how to reset a password or assign a new PIN to a Blackberry (for those thinking we are discussing the prehistory of technology, unfortunately, you are right. I worked on the help desk quite some time ago). After this onboarding, knowledge articles became my best friend. Having worked on many systems, I can confidently say that SF has resolved the issue of suggesting the best articles among the available solutions. But wait, do you even know what Knowledge Articles are? They are support texts for agents, suggesting solutions to a scenario problem. They enable Agents to assist their clients even if they are unfamiliar with the solution to a particular problem.

Knowledge articles in Service Cloud were designed with Agents in mind. But first, make sure to enable it. Open your favorite **Setup**, type `Knowledge` in the **Quick Find** box, and toggle the switch next to **Enable Lightning Knowledge**. Here you can set all the settings regarding SF Knowledge. Every Knowledge Article must be meticulously described during creation. The formula looks like that in the following screenshot:

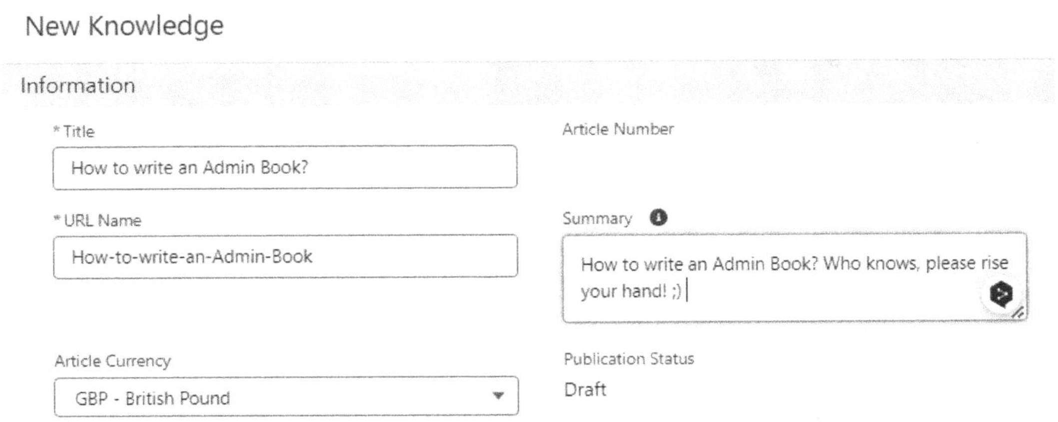

New Knowledge

Information

* Title

How to write an Admin Book?

Article Number

* URL Name

How-to-write-an-Admin-Book

Summary ⓘ

How to write an Admin Book? Who knows, please rise your hand! ;)

Article Currency

GBP - British Pound ▼

Publication Status

Draft

Figure 10.5: Knowledge Article Information

You will find the article title, a URL that is automatically created after entering the title, a brief summary of the article, and interesting features such as **Article Currency**. The last field is just where you can set the current status of the article. As you can see, it is a draft, because until the article is finished, it is best if it remains inaccessible to users. Only after completing the article can you click the **Publish** button, which is located in the button section on the Article record. The following screenshot shows where to enter the article content.

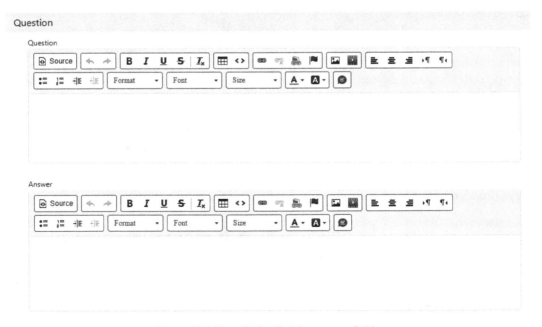

Figure 10.6: Knowledge Article content fields

In the available editor options, you can utilize all the possibilities to convey content clearly by adding images, and links, and emphasizing the most important text elements.

> TIP
>
> If you do not see all the fields on the object on your computer that are shown in the preceding screenshot, then go into the object and check the page layout. Find the field you are missing, add it to the page layout, and save. This way, we can straightforwardly synchronize the objects.

Case management

Assigning cases is a crucial aspect of managing issues within a company. In a service desk department, the structure often consists of agents and team leads, with the service desk divided into specific departments with various specializations such as finance, technical issues, and client contracts. Imagine a scenario where a finance-related case is assigned to someone without financial knowledge. This could lead to bad customer feedback, unresolved issues, and usually escalations affecting the agent's perception. Service Cloud comes to the rescue here. Let us start with the basics of ownership, which I briefly mentioned earlier. Ownership is the responsibility for a specific case. Incoming cases cannot wait indefinitely for someone to accept them, leading to two scenarios:

- The team leader takes care of the assignment, which can cause delays as even leads need breaks
- Service Cloud handles the assignment without needing breaks

What to use for case assignment? Queues can be a starting point. Queues in Service Cloud have been offered as a solution by SF from the beginning. Their role is to manage cases, tasks, work orders, or other records that will be resolved by the customer service team or another department. Queues allow for not having to decide which agent to assign a case to; instead, cases can be assigned to a queue to which a specialized group of agents was previously assigned. Some undeniable benefits of having queues in Service Cloud include the following:

- Grouping records allows for adding cases to a specific queue without going into detail, letting a team leader assign a record to a queue where agents can then thoroughly review the case and determine its priority, the type of inquiry, and so on

- **Case assignment**: Options in queues include manual assignment, where a case can be assigned by a system user, or automatically using automatic case assignment rules

- **Increased efficiency**: Queues allow for the quick identification and assignment of tasks, speeding up case management and enhancing team productivity

The following is a quick guide to creating your queue:

1. Start where you always do, by entering Setup or Service Setup (either will work).

2. Go to your favorite spot – the Quick Find Box, and type in Queue.

3. The system will search for queues and highlight the found setting. Click on the search result.

4. Then press the New button on the page.

5. Now the fun begins. A page with all the settings related to queues will open, looking like the following screenshot.

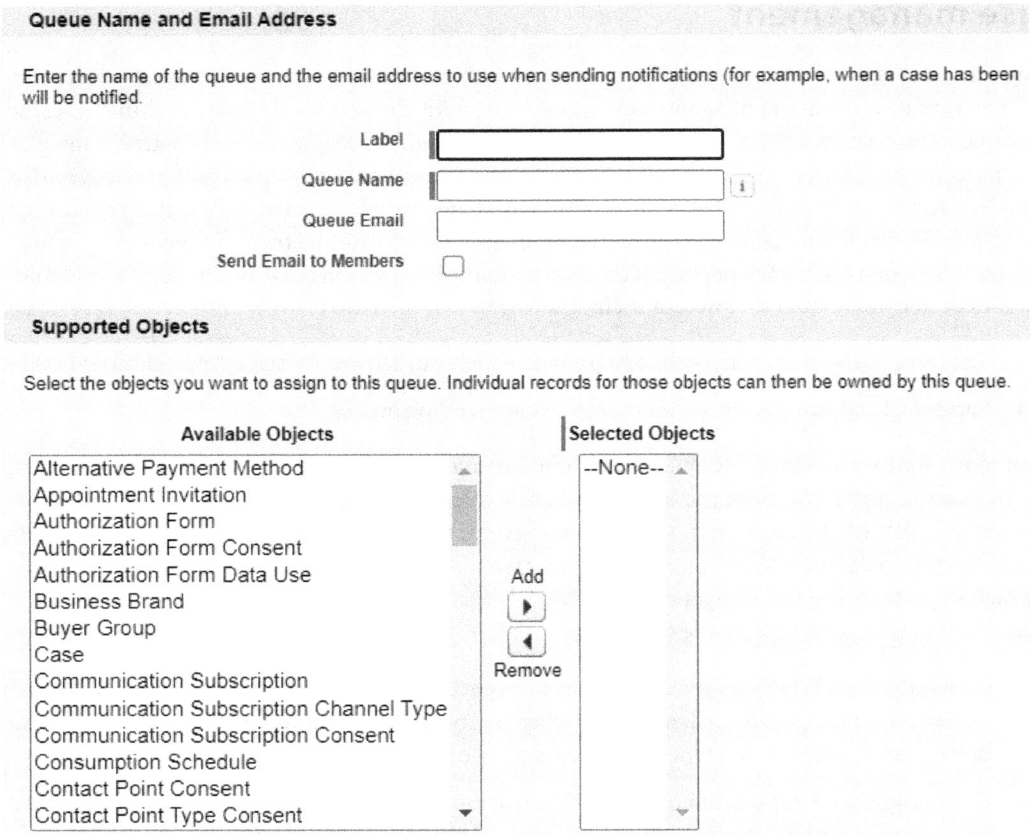

Queue Name and Email Address

Enter the name of the queue and the email address to use when sending notifications (for example, when a case has been will be notified.

Label	
Queue Name	
Queue Email	
Send Email to Members	☐

Supported Objects

Select the objects you want to assign to this queue. Individual records for those objects can then be owned by this queue.

Available Objects		Selected Objects
Alternative Payment Method		--None--
Appointment Invitation		
Authorization Form		
Authorization Form Consent		
Authorization Form Data Use	Add	
Business Brand	▶	
Buyer Group	◀	
Case		
Communication Subscription	Remove	
Communication Subscription Channel Type		
Communication Subscription Consent		
Consumption Schedule		
Contact Point Consent		
Contact Point Type Consent		

Figure 10.7: Queue settings

6. At the start, we have the Label field, where we set the label for our queue. Next are Queue Name and Queue Email. Email addresses are necessary to inform users assigned to the queue that a new case has arrived for them. The last option is a checkbox to activate the sending of informational emails to users.

7. In the next part of the page, we have Supported Objects. Here, we set which records can be assigned to a particular queue. As you can see, not all objects are available; Account and Contact will not be found here, but objects like Case, Invoice, or Order will be.

8. The last setting involves assigning Users, Public Groups, Roles and Roles and Internal Subordinates to the queue, as in the following screenshot.

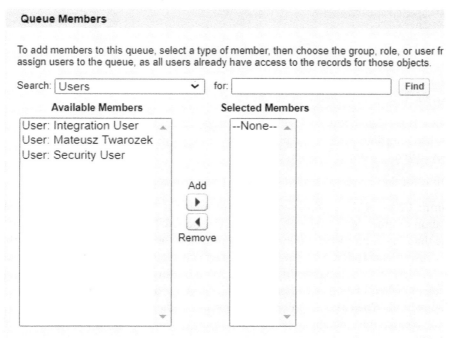

Figure 10.8: Queue members

As you can see, creating queues is extremely simple. So, if you are working with Service Cloud and your help desk colleagues do not have this solution implemented, suggest making a change.

Another tool provided for Service Cloud is **Case Assignment Rules**. These are rules where we establish specific conditions that trigger the assignment of a case to a queue or an agent, looking exactly like the following screenshot.

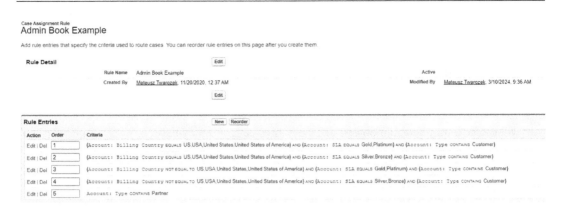

Figure 10.9: Assignment Rules

In them, you set Rule Entries, where you establish the correct order of conditions that are checked sequentially, and if they meet their criteria, they trigger an action.

A twin process is Escalation Rules, which use certain conditions to check values related to customer service, case handling time, priority, and many others—if any of these values deviate from the standard, the case is escalated to a selected person or queue.

And now, time for the superhero. Is it a bird? Is it a plane? No, it is OMNI-CHANNEL!

What is Omni-Channel? It is a tool that allows for integrated customer service across various communication channels, ensuring high fluidity of interaction. Nowadays, help desks have a much harder time than in the past when phone, email, and chat were the only communication channels available to customers. Today, customers can contact call centers through social media, various messengers, emails, phones, and ticketing systems. All of these can be managed by Omni-Channel. For some reason, when I first heard about this tool, an image of a hovering ball with many tubes appeared in my head. When a communication entered one of the tubes, the software inside managed to capture the content information and matched it to the appropriate agent with the relevant skills and importantly, an available agent.

Omni-Channel operates in real time. Cases are assigned without delay to the appropriate agents. But Omni-Channel itself has a few cool super gadgets, as befits a superhero:

- **Skills-based routing**: This is routing based on appropriate skills. It allows cases to be assigned to agents trained to resolve them. Using this part of the configuration, we can be sure that a technical person will not be answering questions related to the latest sales offer from the company.

  Example: A company offering server hosting and domain purchases has many specialists dedicated to various tasks – troubleshooting server or domain issues, collections, finance, and a technical department. Thanks to skills-based routing, new reports are automatically assigned to the person who has the required range of skills required. This results in quick help for the customer and good feedback.

To create such skills-based routing, you must first create skills, skill sets, and mapping, as in the following screenshot:

Figure 10.10: Skills mapping in Omni-Channel

- **Direct-to-Agent**: This feature allows for the direct routing of cases to a specific agent. Often, clients have their favorite agents or a particular agent is resolving a long-standing issue. Instead of transferring all the details to another agent, it's sufficient to switch it directly.

    Example: A customer has an issue with their smartphone turning off constantly, preventing them from finishing a conversation. Borrowing another phone, the customer calls back but does not want to explain everything from the beginning. Thanks to Direct-to-Agent routing, the customer can immediately be transferred to the agent they were previously speaking with, saving time and money (for both the customer and the agent).

- **Secondary Routing Priority**: This feature assesses the priorities of cases. In the event of two identical cases, it manages their assignment based on specific guidelines, eliminating random elements.

    Example: There is an outage! Phones are ringing off the hook, and email inboxes are full. The company sets Secondary Routing Priority so that all calls related to the outage are prioritized. Even though other cases have the same priority, emergency calls are directed to agents as quickly as possible.

- **Status-Based Capacity Model**: This function meticulously monitors the capacity and availability of agents. During configuration, you can assign a maximum capacity for each agent, ensuring agents typically handle an equal number of cases. The status indicates when they're ready to take on a new case. This feature allows for flexible and realistic loading of new cases onto agents, ensuring they are not overwhelmed.

    Example: Agent John is a highly specialized agent with knowledge and multitasking skills. However, his multitasking only allows him to chat with two customers simultaneously and have one phone conversation during this. The system considers the agent's availability and capacity. After checking, it knows that John is on a phone call, so no additional chats will be assigned to him.

As you can see, Omni-Channel is somewhat of a hero for help desks. Of course, it will not cure all ills, and we will probably still listen to the music set while holding a call with an agent. But it will be a significantly shorter wait than it would be without Omni-Channel.

And indeed, time is an incredibly crucial element in customer service. Were you ever frustrated when your case was processed at the maximum time or even longer? You did not know what was happening with your inquiry or whether you would get a response today, tomorrow, or maybe a month from now. That is exactly what **Service-Level Agreements (SLAs)** are responsible for. They are guidelines that are closely monitored because exceeding them results in escalation to a higher level (something an agent doesn't want). An SLA may include time or specific behaviors/actions. Service Cloud aids us with two features in this area – Entitlements and Milestones.

Entitlements specify the conditions a company sets in customer service. The rights and level of customer service contained within are closely monitored. These conditions can be related to the time to resolve an issue, access to support (e.g., 24/7), and many others.

To illustrate this further – a company sells software and has both individual customers and corporate clients. Individual customers are important to them, but they only generate 1/10th of the revenue of corporate clients. Corporate clients are thus extremely important, so for them, support is available 24/7, and all cases must be resolved within 2 hours. This is unlike the conditions for individual customers, for whom agents are only available 5 days a week, and the resolution time is one business day.

To activate the entitlement process, you need to go to **Setup** and enter `Entitlement` in the **Quick Find** box, then find **Entitlement Settings**. The resulting screen is shown in the following screenshot.

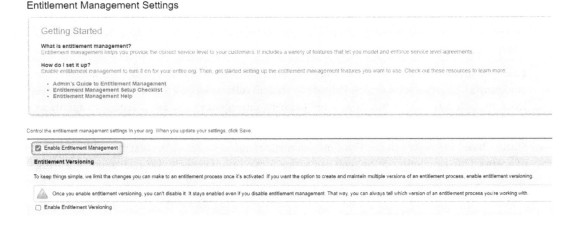

Figure 10.11: Entitlement Management Activation

It is crucial to enable Entitlement Management. This will activate additional configurable features such as Entitlement Processes, Entitlement Templates, and Milestones. And it is Milestones that you are about to read about now.

Milestones are goals that must be achieved to maintain the SLA at the required level. They pertain to actions such as the first response to a case, the time to resolve an issue, or the time required for a case to reach a certain status. Creating a Milestone is straightforward: **Setup | Milestones | New Milestone**. However, their configuration is a bit more complex. This is because you configure Milestones only after linking them with an Entitlement process. The connection looks as shown in the following screenshot.

Figure 10.12: Entitlement Process Timeline

Here we have the timeline of the entire Entitlement Process. By adding a Milestone, you set the minutes to complete the Milestone, the start time criteria, the criteria to be met, and the business hours.

And what happens when the SLA is exceeded? Things like this can also be configured. After exceeding the SLA, a case may be assigned, for example, to another team or escalated to a manager. It all depends on your setup.

Consider the following example of milestone usage: A company sets conditions for priority customers (a checkbox on the Contact object); after creating a case, an agent must contact the customer within 2 hours, and the case must be resolved within 6 hours of the call. The system monitors the agent's actions and if tasks are not completed within the specified time, the system triggers a notification to their manager and sends an escalation.

These two tools streamline agents' work and also improve customer service. Thanks to them, customers no longer wait longer for certain actions, ensuring quality and assuring them that their problem will be resolved.

Now let's talk a bit about communication channels in Service Cloud. Service Cloud is very open to all communication channels. Salesforce always keeps up with the times and responds to any changes and quickly changes itself, such as with Einstein GPT, which was renamed to Einstein Copilot (as an interesting fact, by the time you read this, the name will probably have changed a few more times. Is it already maybe Einstein Gemini?). Therefore, as other ways to contact companies appeared in the world besides just phone, email, and chat, Salesforce immediately added these elements to Service Cloud. And from that moment, lying in bed, you could contact the company via WhatsApp or Messenger. But let us start with the basics:

- **Email-to-Case**: An agent logs in to their company email, copies the entire content, but makes a mistake in the customer's surname, turning Mr. Honkey into Mr. Donkey. No!!! Of course, that's already prehistory. Now, thanks to the email-to-case functionality, a case is created automatically. To configure email to case, you need to go through a few interesting steps:

    I.   Enter **Setup** and search for `Email to Case`, then click on this option from the results.

    II. Find the option to enable email to case by selecting **Enable Email-to-Case**.

    III. But you need to be careful, there's a warning reading **After you enable Email-to-Case, you can't disable it, but you can update its settings**.

    IV. Then save the settings. After reloading the page, scroll to the bottom, and in the Routing Addresses section, add a new email address. While adding this, you can also set various options such as case priority and whether a task should be created, among others.

    V.   After saving the address, verify it by pressing Verify next to the email. Then, you will receive an email from SF with a verification link.

    VI. Click and verify the address. Set up forwarding or a transport rule on your mail server so that all emails sent to the email address (set in the previous step) are automatically forwarded to the Salesforce email address. So, if someone sends you an email at `imNotRealEmailAdress@gmail.com`, the message will be redirected to the Salesforce-provided email address.

    VII. After setting up forwarding, only the final touches remain and the configuration is ready.

- The next channel is telephone, the most basic communication channel. But of course, we do not connect it in the standard way. Instead, we use **CTI**, or **Computer-Telephony Integration**. This refers to third-party solutions that allow the receiving and making of telephone calls, recording them, listening to them, and creating the right Call Flow, which guides the customer through various paths such as choosing the topic, a welcome message, connecting with the customer, or asking them to leave a message. Most companies offer ready-made solutions and include Professional Services, i.e., people who will configure such a contact center. Notable solutions include the Natterbox and Vonage software.

- Next is live chat. This solution is very commonly used. When creating live chat, a code is generated that needs to be placed on the website. This allows customers visiting the website to freely write to your support. This solution offers a range of extensions. There are even a few companies worldwide that offer real-time translation. This type of solution requires developer work, but in this case, it is worth the price. Companies such as Fluentic, Language I/O, or 1440.io can boast of such solutions. We can also easily add a chatbot to our live chat that will verify the customer's problem, suggest certain solutions, and then redirect the customer to the appropriate agent. When configuring a chatbot, remember that we are not talking about something that has internal AI and can solve the customer's problem on its own. We are talking about a path that we configure, through which the customer is directed to the final point in a simple and pleasant way. Thanks to the configurable paths prepared by SF, we do not need a developer to create scenarios. And most importantly, our digital buddy is available to the customer 24/7.

- Now, something more modern – social media. I remember when Facebook appeared in my life, and I announced every love success and failure to the world in the form of emo poetry (don't judge, we all make mistakes). From the start, it had huge potential. And Salesforce decided to utilize this potential. Thanks to connections, customers can now contact companies through most social media platforms and messengers including Messenger, WhatsApp, Viber, Signal, and Telegram. Of course, we can also find companies in the market that already offer such a solution, such as the 360 SMS App, which includes connectors to a large number of internet messengers.

> **TIP**
>
> If you start a project that will be a significant undertaking, such as a long project of connecting internet messengers to Service Cloud, consider whether you want a ready-made application versus a custom solution.
>
> If you decide on a custom solution, it is a larger one-time expense, but such a solution is permanent. With the ready-made solutions offered by various manufacturers, remember that they are usually subscriptions that you pay in annual or monthly fees. However, if the client has their own developers building the solution for them, that's fine, but if it is outsourced to a partner, it's worth considering which will be better, because a ready-made application developed by the manufacturer will always be up to date. A custom solution needs to be updated. The choice is yours.

- The next communication channel is a self-service portal. This solution is usually suggested by architects when the company already has an external database set up and wants to give clients access to it. Theoretically, this should significantly reduce call-center traffic. But that's just theory: in reality, only 1 in 10 clients will enter the portal and look for a solution themselves. These types of portals are built on Experience Cloud. When building the page, you can even find a fully available template that only needs to be adapted to your company. Companies often throw in the famous FAQs, thanks to which one in ten of your clients will find the answer to their pressing questions.

Now that we have our communication channels set up, people assigned to queues, and redirections configured, what we need are reliable KPIs and the tools to monitor them, namely reports and dashboards. When I worked on a help desk, each of us had our own dashboard with data covering our cases, long-standing issues, the queue, average response times, and many other useful pieces of information. Remember, Salesforce is a tool where we first build the framework, then refine it, and once everything is ready, we can extract comprehensive information from it. Therefore, I want you to learn about the reports that can transform the life of any help desk. Here are some examples of such reports:

- **Average response time for a case**: This report shows the average response time. It allows managers to monitor SLA compliance in resolving cases.

- **Case types by percentage**: A report showing the types of cases coming from customers. You can also apply different time periods here, such as this month, last month, and this year, allowing you to see trends in the arrival of specific types of cases and when it's worth having a larger team to assist in resolving cases.

- **Customer satisfaction level – the famous CSAT**: If your company sends out CSAT, aka customer satisfaction surveys, and such ratings return to Salesforce, that's great because you can easily export it in a report and compare which cases have the lowest ratings and then identify the cause of the low ratings. Sometimes it may require additional training for an agent, supporting them with additional tools, or sometimes just reviewing the product.

- **Cases by Channel**: This report provides an overview of which channel most cases from customers are coming from. With this information, the manager can assign additional people to manage the chat if it turns out to be the most popular, for example.

If we already have reports sorted, we should next think about dashboards. Here are some I can recommend that will enhance the work experience in Service Cloud:

- **Agent efficiency**: This can include data such as the number of cases resolved by department, average case resolution time, or the CSAT index.

- **Query trends**: This dashboard can provide extremely crucial data on trends in certain periods. This way, managers can foresee the need for additional support, just like with reports – but here in a more comprehensive way.

- **KPIs**: Team efficiency, customer satisfaction, and areas needing attention – remember that leadership wants specifics.

- **SLA monitoring**: A dashboard showing cases that might soon exceed the SLA is very useful. This dashboard will prevent deadlines from being crossed, thereby eliminating the risk of case escalation.

As you can see, reports and dashboards can provide extremely important data for managers and company executives. With them, not only can the customer service process be optimized, but adjustments can be made to annual or even monthly trends, which will definitely impact the final CSAT ratings.

"But wait, cases are also available in Sales Cloud, so why do I need a Service Cloud license?", you might ask. And that is a very good question (yes, I realize I am talking to myself).

In Sales Cloud, you have access to very basic case management. That means you can log data on cases, then track them, insert comments, and by transitioning from one status to another – close a case. These cases contain basic information and offer an overview to the sales team.

However, with a Service Cloud license, you can build a real Godzilla of help desks, a defender of customer service, and a warrior for good CSAT. Okay, I got a bit excited. But it is all true.

A Service Cloud license offers advanced case management functions such as assignment rules, tracking, and case escalations. Then there is our helper, Omni-Channel routing, and support for multiple communication channels.

So, it all depends on your needs. If your company is in the **Small and Medium-sized Business (SMB)** sector, you might try setting up a help desk on the standard Salesforce license in Sales Cloud. However, if the functionalities I described in this chapter will be useful to you, I definitely recommend you and your company purchase such licenses.

What does the future of Service Cloud look like? I think we can all play fortune-tellers with a crystal ball and predict something very similar. Two letters – AI. A big step will be training AI on our procedures and introducing it as a fully functional agent to handle customers via chat. Another interesting solution, which will also be a huge technological leap, is real-time translation during phone calls – imagine being an agent who can speak every language in the world, thus not being disqualified from a job for not knowing Klingon. As you probably know, Salesforce always keeps up with the times and introduces new technologies. So, the coming years are very exciting, and the AI revolution will certainly bring a lot of innovations.

Summary

And with that, you have learned about the structure of Service Cloud, why Salesforce released such a product, and its functionalities that can help in customer service. I am sure that for many help desks, Service Cloud has changed customer service by 180 degrees. Thanks to its functionalities and simplifications, agents can breathe easier and solve cases much more comfortably. Management can finally extract relevant data from the system and improve the quality of service given to their customers. But wait, if we already have a help desk, it is good if that help desk has someone to serve and to have customers, we need to sell or offer something. And here comes the very popular Sales Cloud, which will be discussed in the next chapter. See you in a bit!

11
Sales Cloud

In this chapter, I would like to describe to you the most important, even core, functionality of Salesforce, known by the name Sales Cloud. This is, indeed, the cloud that sells the best; it is the most popular, and you, dear reader, will most likely encounter it on your journey, whether as a user or an administrator. In other words, Sales Cloud is the alpha and omega of the Salesforce world, something that Mark Benioff, like Prometheus stealing fire from the gods, took from the gods – not the ancient gods of Olympus but the more modern gods of the internet. Without Sales Cloud, Salesforce would not be as recognizable, and probably without Sales Cloud, there wouldn't be other clouds such as Service, Marketing, and Data.

So, is this mythical Sales Cloud real, as we've already mentioned? Yes, it is! What is Sales Cloud then? As Salesforce describes, Sales Cloud *"unlocks efficiency by managing Contacts, Leads, Opportunities, and customer Accounts in one place."* Salesforce Sales Cloud is thus a CRM in its purest form, but with one important note: it is often rated by many as the best CRM in the world! Sales Cloud establishes the groundwork for boosting revenue by employing Salesforce automation, accelerating the closure of deals, and anticipating forthcoming sales by gaining insight into pipelines and forecasts.

How is this all even possible? What cosmic features are included within Sales Cloud? Simply put, it can be said that Sales Cloud consists of four tabs in Salesforce: **Leads**, **Accounts**, **Contacts**, and **Opportunities**. This is, of course, a significant simplification, so let's add to them **Campaigns** and **Orders**. Still a substantial simplification, so I'll also mention **Activities** and **Quotes**. You see where I'm going with this. Contrary to appearances, Sales Cloud is not that simple, and there is more to it than just your database of companies and contacts. Sales Cloud is, as mentioned, not only about Accounts and Contacts. It is true that most users primarily use these tabs in Salesforce, but beyond them, there are numerous features of Sales Cloud, some of which may be used less frequently or not by everyone, but they are an integral part of the system. Thanks to these features, Sales Cloud enables easier, faster, more transparent, and automated business management. Adding the recently trendy AI to the mix, we have a recipe for successful sales management software. Okay, enough talking. In this chapter, I would like to examine with you the key functionalities of Salesforce Sales Cloud and the associated functionalities. In this chapter, we will focus on the following:

- Seller Home
- Leads

- Accounts

- Contacts

- Activities

- Campaigns

- Opportunities

- Quotes

- Forecast

Let's start with where every Salesforce user is likely to land after logging in – Seller Home.

Seller Home

Seller Home is a relatively new solution from Salesforce designed to theoretically facilitate the work of sales professionals operating within Salesforce. It represents a natural extension of the classic **Home** tab available to all users, with the addition of certain elements tailored to the needs of sales teams. Is it a perfect solution? Not entirely. However, it is a step in the right direction, and it's encouraging that Salesforce has recognized the need to customize certain views for what is arguably their largest user base – sales departments.

Seller Home includes several interesting elements, some of which are entirely new, while others are derived from the former default **Home** tab accessible to all users. The overall presentation is quite intriguing, as can be seen in the following screenshot.

Figure 11.1: Seller Home

Due to the presence of some new features, I will briefly discuss what is available on Seller Home:

- **Close Deals**: This section displays Opportunities owned by the user that are scheduled to close in the current quarter. It assists the user in effectively managing their sales pipeline.

- **Plan My Accounts**: Here, the user can view their owned Accounts. Additionally, it provides information on the planned and completed activities associated with these Accounts.

- **Build Relationships**: This area showcases the user's Contacts created in the last 90 days, offering easy access to the newest Contacts. The user can also view the number of activities related to these Contacts, aiding in the planning of interactions to establish and maintain necessary relationships.

- **Build Pipeline**: In this view, the user can see Leads owned by them and created in the last 30 days – relatively fresh Leads. It also displays the number of activities associated with these Leads, as engaged Leads have the potential to transform into new Opportunities, contributing to a robust pipeline.

- **My Goals**: This feature is quite interesting. Users can input weekly or monthly goals for meetings, calls, or emails. After setting the goals, the view is automatically updated as new meetings are created, calls are logged, or emails are sent. This functionality helps in tracking progress related to these activities. The following screenshot illustrates how to set up goals and what the view of completed activities looks like.

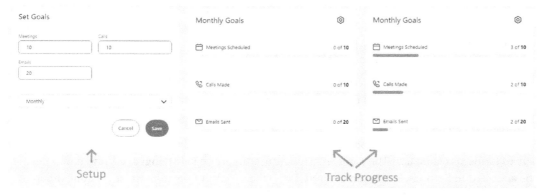

Figure 11.2: Seller Home Goals tracking

- **Today's Events**: This is a very basic view of meetings planned for today.

- **Today's Tasks**: This is similar to **Today's Events** but shows Salesforce Tasks, for example, calls.

- **Recent Records** – Here is a list of recently opened records. It helps you access records that you visited recently faster.

As you can see, Seller Home serves as the central hub for sales professionals within Salesforce. It provides a comprehensive overview of essential information and tools to streamline the sales process. Users can access important features and data related to their sales activities right from Seller Home.

From Seller Home, users can navigate to different sections of Sales Cloud, such as **Leads**, **Accounts**, **Contacts**, and **Opportunities**, to dive deeper into specific aspects of their sales workflow. In the next section of this chapter, we will go through all these mentioned objects, starting with Leads. Let's go!

Leads

We briefly discussed Leads in *Chapter 1, Introduction and Getting Started with Salesforce*, of this book; however, in this section, I would like to expand on this description and show you more. As a reminder, in Salesforce, a Lead is a place where potential, preliminary sales opportunities can be recorded. A Lead is a combination of fields from both Account and Contact, as it contains information such as the potential customer's first name, last name, phone number, and email address, but also the name or address of the company they work for. Alright, but where can I get such Leads? That's a good question because unless you're Harry Potter and can conjure them up yourself, they won't magically appear in your system. Let's list the most common sources of Leads in this place. These include the following:

- **The salesperson's own efforts**: The salesperson actively explores the market, contacts potential customers, and enters them into the system as Lead records.

- **Website form**: You may have noticed that some companies place contact forms on their websites. After filling out such a form, your inquiry is transmitted to the company. Salesforce Sales Cloud provides dedicated functionality called Web-to-Lead, which allows the creation of such forms. We will describe this functionality in more detail later in this chapter.

- **External database purchases**: Salesforce can be enriched with data from external databases. Some companies legally sell databases that can be integrated with Salesforce.

Salesforce has a dedicated tab for browsing and managing leads. Of course, you can use standard list views to see them, but what's interesting is that Salesforce recently created a new view dedicated solely to Leads and Contacts. The name of this new user interface is **Intelligence View**, and it needs to be activated in **Setup** before it can be used. Let's take a look at how it looks, as it's quite different from the standard list view.

Figure 11.3: Lead Intelligence View

As you can see in the preceding screenshot, Salesforce has made a significant improvement by upgrading the standard list view with additional information that strongly correlates Lead (or Contact) records with activities. This is a good direction because, as we know, a salesperson's effectiveness (i.e., someone who works on Leads) largely depends on their activity level, such as the number of conversations conducted (and their quality, of course), when such activity took place, and whether any Leads are uncontacted. All of this is provided to the user through the Lead **Intelligence View**. As you can see in the screenshots, you can filter Leads that have no activity (**No Activity**), have no upcoming activities (**No Upcoming**), or had a past activity planned but not completed (**Idle**). You can also quickly find Leads that have an activity planned for today (**Due Today**). Additionally, by hovering over a Lead record and clicking on a special icon on the right side, a drop-down window appears showing the activity history for that Lead. From this view (and also from the Intelligence View list itself), you can also create new activities, such as Tasks, Events, or Emails.

As you can see, Salesforce learns from user feedback and tries to improve its solution so that working with record lists is even faster and, above all, more optimal and user-friendly. Standard list views have been around for a while, and it would indeed be beneficial to refresh them; in my opinion, the Intelligence View is a step in the right direction. Of course, I also think this is only a small step, as this view currently has its flaws (which I will overlook for now) and is only available for Leads and Contacts at the moment. However, as the saying goes, "Rome wasn't built in a day," so I am eagerly awaiting Salesforce's next steps toward improving the user interface for record lists. In the meantime, let's return to the view of the Lead record itself, which hasn't changed for some time.

An example of a Lead is shown in the following screenshot. In this case, the Lead source has been identified as **Web**.

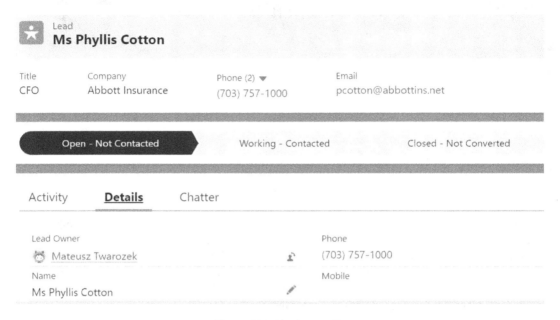

Figure 11.4: Lead example

Now that you know where Leads can come from, it's time to learn what you can do with them. But before we dive into that, let me pose a riddle to you! Do you know the difference between a Lead and a Contact? Or are they perhaps the same thing? Exactly! The answer is not straightforward; it's a rather tricky question. If you're just starting your Salesforce journey, concepts such as Lead, Contact, and Account might get a bit confusing. Don't worry; they can still be confusing even with more experience on the platform. This is because every Lead can also be a Contact, and vice versa. Let's see how it actually works in the system.

Imagine you have a customer, a company called Super Bus. Mr. Edward Stone works at Super Bus. Since Super Bus is your customer, Mr. Edward Stone is recorded in Salesforce as a Contact. He's simply a Contact linked to the Super Bus Account. But at the same time, instead of calling you, Mr. Stone filled out the contact form on your company's website, which is connected to Salesforce. What happens then? Mr. Stone enters Salesforce as a Lead. In this way, you will have Mr. Edward Stone in Salesforce twice – once as a Lead and once as a Contact. Is it useful? Yes! Is it troublesome? Also yes!

Let's start with the positives. Sometimes it's good that every contact through the website, meaning those who fill out a form, creates a new Lead. This is because such forms may relate to various products you sell, and handling these Lead inquiries can be taken care of by different dedicated individuals, not necessarily you right away. If your company offers a wide range of services, it might be the case that your customer is not only your customer – someone else from your company might be selling them something (else). Simultaneously, Salesforce, by separating Leads from Contacts, Accounts, and Opportunities, allows you to register the initial stages of sales or, more precisely, sales potential. Leads are objects in Salesforce that often have many records, but generally, a small percentage of them transform into actual sales. It strongly depends on what and to whom you are selling, but generally, salespeople work with Leads, closing more of them than converting them into Opportunities. They contact many Leads, verify their interest in your service or product, mark their activities through tasks or events, and change the status of leads. But ultimately, working with a Lead can end in only two ways: unqualified or converted. An example of a Lead marked as **Closed - Not Converted**, so unqualified, is shown in the following screenshot.

Figure 11.5: Lead unqualified

While *unqualified* is self-explanatory and indicates that we won't be further processing a particular Lead, the term *conversion* sounds quite mysterious. The magical words *Lead conversion* are frequently heard in the Salesforce world. What does it mean? What is this conversion, who does it, and why? It's very simple. If, after verifying a Lead (for example, through a phone call with that person), we see that successful sales are possible, Salesforce allows us to move that Lead to a higher level. In Salesforce, similar to collecting levels in a Mario game, converting a Lead is like moving a potential customer to a higher level.

During the Lead conversion in Salesforce, it creates three records: Account, Contact, and Opportunity. It is evident that in the sales dimension, Lead conversion in Salesforce is a significant matter because it results in the creation of an Opportunity (which, of course, can be omitted as it is not mandatory). The salesperson aims to successfully close this Opportunity within a specific timeframe, and often there is a target for the number of such acquired customers. Since I would like you to see for yourselves how Lead conversion looks, we will practice it together in a simple task:

1. Open any Lead record.

2. Locate the **Convert** button and click on it.

3. Check the data and objects to which the Lead will be converted. You should have a view similar to the following screenshot:

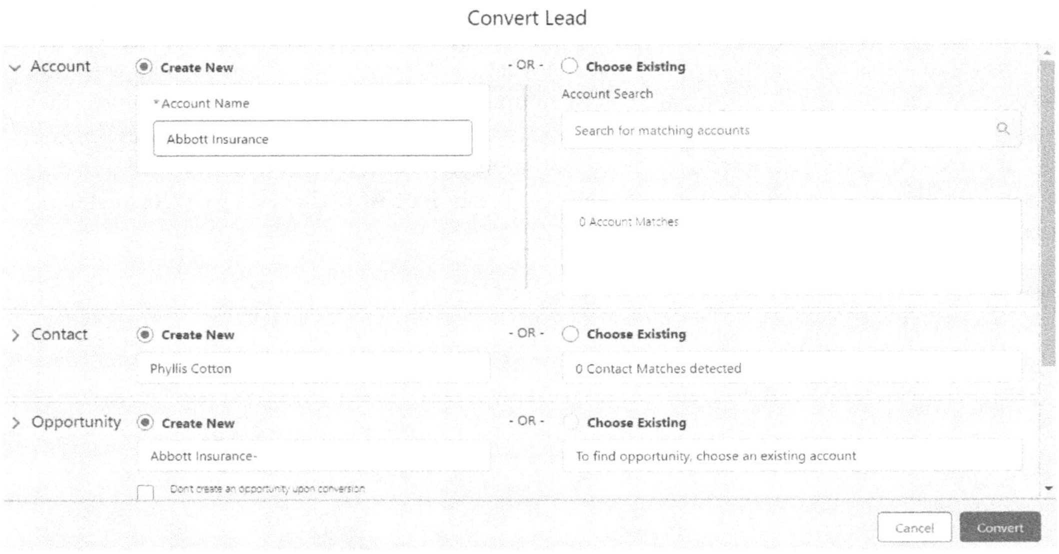

Figure 11.6: Lead conversion page

4. Press the **Convert** button.
5. Access the records that were created after converting the Lead, such as Opportunities, to check what was created after Lead conversion.

Alright, so now you know everything essential about Salesforce Leads. You're aware that a Lead is a crucial record in Salesforce Sales Cloud, often marking the beginning of a user's journey. You understand that a Lead can either be closed or further processed by converting it into other objects. One of these objects is an Account, and that's what we'll focus on in the next part of this chapter.

Accounts

I think, in this case, there's no need to delve too deeply. An Account in Salesforce contains the company data you input. These can be your clients, potential clients, partners, or even competitors (to mark this, you can use the standard **Type** field). Importantly, Accounts are related to Contacts and Opportunities. This means that by accessing the profile of a specific company, you can immediately see which Contacts belong to that company or view historical and current sales opportunities.

Account records also include the company's contact information, tax numbers, and so on. These are rather static data, making this object less frequently edited and, therefore, less attractive to users. However, it exists because there needs to be a common part connecting Contacts and Opportunities. Yet, it doesn't always have to be this way. One thing I often do to make Account records more appealing is to use fields such as Rollup Summary, summing up the total sales amounts for a particular company. This adds dynamics to Account records and provides users with more data. However, remember to consider data security when planning such solutions, as not everyone in your company with access to Accounts might need to see such details.

However, Accounts alone don't constitute the whole without connected Contacts. Let's see how this Salesforce feature works.

Contacts

In this case, much like with Accounts, there aren't too many mysteries. Contacts are simply records, most often associated with Accounts, meaning employees of a particular company or individuals with whom you establish cooperation during sales activities, such as writing to them or calling them if you are a salesperson. Contacts appear in Salesforce after converting Leads or can be created directly on Accounts by clicking the **New** button in the Account Related List.

The Contact object includes fields such as first and last name, email address, one or more phone numbers, and address details. These details may differ from the registration data of the associated Account, as the contact person might work in a different department or city. You can record activities related to the Contact, such as calls or meetings. Additionally, you can add notes about important matters.

An important Salesforce functionality related to Contacts is the ability to connect one Contact to multiple Accounts. Sometimes, one person may be associated with several companies simultaneously, and we'd like to note such a fact. This could be a recruiter recruiting for several subsidiaries of the same parent company or a chief financial officer overseeing multiple companies.

Salesforce allows us to mark such dependencies, utilizing an intermediary object that enables us to link one Contact with one, two, or multiple Accounts. Before this feature was available (and I remember those times!), either Salesforce users had to create duplicate Contacts and link them to different Accounts, leading to unnecessary records in the database, or custom solutions were developed at the request of clients. Fortunately, at some point, Salesforce recognized this issue and created the necessary functionality that meets the most important requirements of users in this dimension. Let's see together how to connect a Contact to more than one Account. The following is a description of how to enable this functionality in Salesforce. There is also a chance that you have it enabled by default.

In that case, skip steps 1-3 and follow along only with the process of linking a Contact to different Accounts:

1. From **Setup**, enter `Account Settings` in the quick find box, and then select **Account Settings**.

2. Select **Allow users to relate a contact to multiple accounts** and click **Save**.

3. Add the **Related Contacts** Related List to the Account user interface.

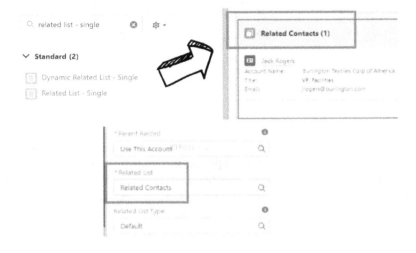

Figure 11.7: Related Contacts

4. Look for a Contact and click the **Add Relationship** button.

Figure 11.8: Related Contacts new relationship

5. Set up the new relationship and save your changes – you may use the Contact that is already on the Account but link it with another Account. In the following example, the **Jack Rogers** Contact is already linked with the **Burlington Textiles Corp of America** Account, but I have also linked it with another Account, **Piramid Construction Inc.**

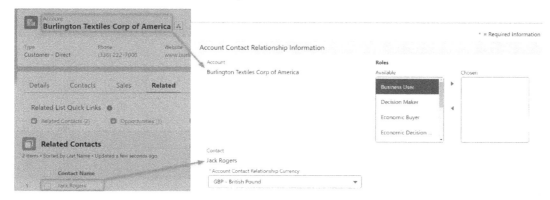

Figure 11.9: Related Contacts new relationship setup

Congratulations! You have just learned how to relate one Contact to multiple Accounts. You may spread this knowledge further. Believe me, your users will be more than happy knowing this is possible in Salesforce.

Activities

Activities are a fundamental feature that virtually every Salesforce Sales Cloud user grapples with. Of course, I used the phrase *"grapples with"* as a joke, but to be honest, activity updates are often a tedious task not always favored by Salesforce users. Of course, Salesforce allows for significant simplification and automation of this process. There is the option to install the Salesforce plugin for Outlook, which allows logging selected emails or events. You can also enable Einstein Activity Capture integration, which automatically associates emails with the relevant records in Salesforce: Contacts or Leads if the email is related to them. While these solutions are not without flaws, users still largely manually add activity records in Salesforce.

How do activities work? In a very straightforward way. In Salesforce, there are three objects associated with activities: Tasks, Events, and Email. They are distinct, so to create a new Task, you use a different button than when creating a new Event or Email. Of course, you can add such activities to any object, even a custom Object (if the object is set up accordingly). Let's take a look at how the **Activity** tab appears on a sample Lead.

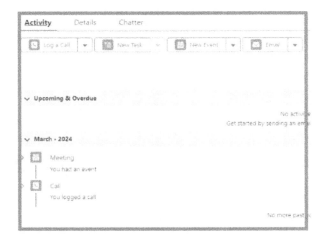

Figure 11.10: Activities Timeline example

As seen in the preceding example, in addition to the buttons mentioned earlier, Salesforce interestingly displays completed or scheduled activities. This is the Activity Timeline, which clearly shows us which activities are already completed and which are still pending, all in the context of a specific record – in this case, the Lead named Mr. Barth Boxer.

Similar activities can be carried out on Contacts, Accounts, or other objects. However, it's worth mentioning that activities such as Email are typically associated with Leads and Contacts. You can see how it looks in the following screenshot.

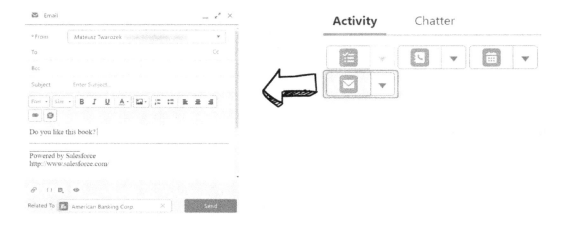

Figure 11.11: Email activity example

As seen in the preceding screenshot, Salesforce allows you to send an email directly from the system. You can enter any content (or use pre-prepared templates) and add additional recipients and attachments. It all looks similar to other email programs, but it takes place within Salesforce, and the email sent from Salesforce is logged as an activity in Salesforce. However, what happens when someone replies to the email? Will it go back to Salesforce? In this case, no. The reply email will go directly into your email inbox. If you want the message to go to Salesforce, you need to explore the additional feature called Einstein Activity Capture or read more about Cases in *Chapter 10, Service Cloud* in this book.

As you can see, Activities are a very interesting object in Salesforce, allowing you to add Tasks, Events, and emails to Salesforce records. It helps organize and plan users' work and, ultimately, accelerates it through integrations with Outlook or the use of the Einstein Activity Capture functionality.

As we mentioned earlier, Activities are one of the features that sales representatives often use. In this book, I would like to also mention other objects that they can use in their work – Campaigns and Campaign Members. Let's discuss them now.

Campaigns and Campaign Members

You could say, *"C'mon, we were supposed to talk about Sales Cloud, and here it seems we will be learning about campaigns?"* Are these marketing campaigns? Yes and... no! In Salesforce Sales Cloud, we have functionalities similar to marketing tools because we have access to the Campaign and Campaign Member functionalities. What are they used for? Imagine that our marketing department is running a campaign targeting our biggest clients. During this campaign, the sales team needs to contact representatives of the largest clients to invite them to an event planned for a few months later. The ultimate result of this contact should include whether the client's representative was contacted, whether they received an invitation and information about who accepted it so that the marketing department can efficiently count the guests and plan the event. The entire event can be planned using Campaigns. In a Campaign, you can name the campaign and specify its duration, budget, expected effectiveness, and so on. Such campaigns are usually planned by the marketing department or sales management. Typically, a set of records that will participate in this campaign, such as Leads, Contacts, or entire Accounts, is also assigned at this stage. However, Leads or Contacts are most commonly used. Once created, the campaign is handed over to the salespeople, who may be tasked with calling their clients to confirm the proposals made to them, such as inviting them to the event, and checking whether the client will accept or decline the invitation. All of this must be marked in Salesforce, and the Campaign Member object is used for this purpose, which is associated with the campaign. In short, each Campaign has many Campaign Members, which can be Leads, Contacts, or Accounts. In our example, we mentioned that the event is organized for our most important clients, so the Campaign Member records would likely be Contact object records. Let's look at the following screenshot to see how such a Campaign could look in practice.

Figure 11.12: Campaign in progress view

In the preceding example, you can see that we have a planned campaign, and currently, there are three Contacts in it. We can also see their statuses and detailed information about the Campaign. However, this view will be more interesting to the campaign creators, that is, the marketing department or sales management. What is important for someone from the sales department is the ability to mark the status of the Campaign Member, that is, the Contact they contacted regarding the organized event. The ability to mark the participation of a Contact in the organized event will appear in the **Campaign History** list, where we can set the current status for such a Contact. In the following example, you can see that we sent an invitation to Mr. Andy Young. We will wait for a response, and if Mr. Young is interested, we will change the status. Thanks to this, the marketing department will see another person for whom to prepare materials, print conference badges, and perform similar activities.

Figure 11.13: Contact included in Campaign as Campaign Member

Campaigns and Campaign Members (so, for example, Contacts shown in the previous screenshot) allow the sales department (but not just them) to effectively collaborate with the marketing department in planning and implementing various campaigns targeting existing or new customers, complementing the sales pipeline with new Opportunities. We will focus more on Opportunities in the next part of this chapter. This is a crucial functionality of Sales Cloud, so we will delve into it more thoroughly.

Opportunities

What would a salesperson be without selling? A deserted island without inhabitants, a sailor without a ship, or the *Friends* sitcom without Joey? Of course, this is a joke, but this illustrates what Sales Cloud would be without the **Opportunities** tab. Opportunities are a key feature of Salesforce Sales Cloud. **Opportunities** is a place where a salesperson looks very often because it is the tab where they plan future sales, which can be carried out for both existing and new customers.

> **Tip**
>
> Sales departments are sometimes divided into separate teams: farmers, individuals who take care of existing clients by upselling new services or products while maintaining relationships, and hunters, salespeople who acquire new orders and sign new contracts. Very often, both teams work with Salesforce Opportunities, but they have different statuses, milestones, or products they sell.

Opportunities themselves are basically a straightforward object, and their records are directly linked to an Account. After creating an Opportunity, you can see simply what Opportunities a specific Account has. There may be more than one if you're selling various products or services to the same customer or if different sales teams or salespeople are handling these transactions. Let's see how to create an Opportunity record in Salesforce:

1. Go to the record of any Account.
2. Click on **New** in the Opportunities Related List.
3. Name the Opportunity and select the close date (for example, three months ahead).
4. Choose the current stage of the Opportunity.
5. Save the Opportunity.

> **Tip**
>
> The stages available when working with Opportunities are related to a feature called Sales Process that you can configure in Salesforce Setup. The Sales Process feature allows you to create different stages for various types of Opportunities. For example, you may want to have different sales stages for companies that are already your clients, where you just want to upsell them, and different ones for companies that are potential new clients. Perhaps you'd like to differentiate stages based on the products you are selling. Thanks to the Sales Process configuration, you will have the ability to do so.

Filling out the fields on Opportunities and changing stages toward **Closed Won** or **Closed Lost** is not the only aspect associated with professional and effective work with Salesforce Opportunities. Working with Opportunities also involves assigning Contacts with whom we communicate on a specific sales matter. Adding Opportunity Products, creating Sales Quotes, or working with the appropriate

Opportunities view to make it the most user-friendly are other elements that are worth a more detailed description. Let's delve into them now.

Opportunity Contact Roles

Opportunity Contact Roles are nothing more than the possibility to mark the role that a Contact plays in relation to the Opportunity we are working on. Usually, we already know with whom we want to do a particular business, and at this stage (Opportunity), it should be clear to the user. Perhaps we are talking to a decision-maker, or maybe a purchaser. In a given Opportunity, we may be in contact with more than one contact person, and they may have different roles. Salesforce Sales Cloud allows us to mark and assign multiple roles to Contacts and Opportunities. Additionally, the list of roles is fully configurable, so you can create your own list to fit the specifics of your company or sales. Once you have a Contact record entered in Salesforce, adding Contact Roles to Opportunities is done in a very simple way, as described in the following steps:

1. Go to the Opportunity that interests you.

2. Locate the **Contact Roles** Related List, which is typically found on the right side of the Opportunity, and click the **Add Contact Roles** button.

3. Choose the Contact or pair of Contacts you want to associate with the Opportunity as Contact Roles. You should see a screen similar to the following one:

Figure 11.14: Adding an Opportunity Contact Role

4. Click the **Next** button.

5. Assign the role to the selected Contact or Contacts, and that's it!

Assigning Contact Roles to Opportunities may sometimes seem like a tedious or even boring task, especially when a salesperson is responsible for numerous Opportunities. However, it's important to remember that this process helps organize and supervise work. Firstly, because working on Opportunities often involves teamwork, and secondly, in the event of any unforeseen circumstances such as a salesperson's absence (due to vacation or changing employers), Contact Roles provide a clear record of with whom the salesperson communicated while closing specific Opportunities.

Opportunity Products and Quotes

Another thing that a salesperson often does when managing Opportunities is adding products (goods or services) they sell to a particular Opportunity. This is possible because Salesforce provides the ability to add Products along with price lists (Product Pricebooks), which can later be used on Opportunities. Adding products is a simple task, and assuming that we previously entered them into Salesforce, it doesn't take much time. However, it is an essential activity because it is through the added products that we know what we are selling, at what price, whether the price is different from the list price, and so on. Okay, but it's time for an example. Let's say we want to sell two products to a selected customer. To demonstrate this, I will use an account created on Salesforce Trailhead. My products will be a diesel engine (a bit of a relic from the past!) and, additionally, a **Service-Level Agreement (SLA)** related to it. Now, let's see how we can add sample products to any Opportunity:

1. Prerequisite: You must have added Products in Salesforce before you can assign them to an Opportunity.

2. Go to the Opportunity you are interested in.

   Locate the **Products** Related List and click the **Add Products** button, as shown in the following screenshot.

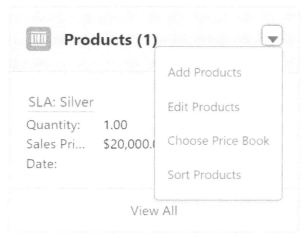

Figure 11.15: Adding Products to an Opportunity

3. Choose one or more products and press the **Next** button.

Figure 11.16: Selecting Products

4. Add a product quantity or edit the price, as shown in the following screenshot, and save your choice.

Figure 11.17: Product quantity and sales price

5. See the Products on the Opportunity. You should have a view similar to the following:

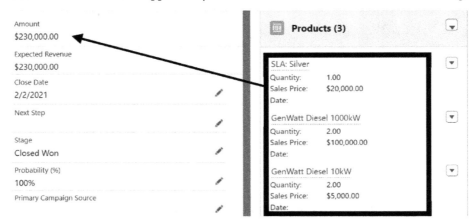

Figure 11.18: Opportunity with Opportunity Products

In the preceding screenshot, you can also see that, in addition to adding Products to the Opportunity, two fields were updated: Amount and Expected Revenue. How is this possible? Well, these two fields depend on the selected products, and the values there are automatically calculated based on the prices of the products, their quantities, or, in the case of Expected Revenue, the probability that the Opportunity will be closed.

As you may have noticed, adding Products to an Opportunity is not particularly difficult; but, of course, we used a very simple case. In practice, quoting often involves various stages, including approval processes, demo presentations, and similar activities. So, remember that this process can be much more complex. Let's talk about one type of complication right away. Imagine that a customer wants not just one but, for example, two quotes for two different types of products that you will propose to them. Of course, you could create two separate Opportunities and add different products there. Sometimes, this solution makes sense, but Salesforce gives us another option to handle such a use case, and it's called Quotes.

Salesforce Quotes are an additional feature in Salesforce that allows you to add various combinations of products or services you want to offer to your customers. The Quotes module allows adding multiple Quotes under one Opportunity, processing Quotes internally, and creating additional approval processes per Quote, so that you can internally confirm these offers before presenting them to the customer. Ultimately, Quotes even allow generating PDF documents in your branding, which you can send to the customer. This requires some additional configuration, but it's not very complicated. The advantage of creating Quotes is the ability to build different configurations of your products or services and the ability to internally approve and check which combinations will be most appealing to your customer. Ultimately, the Quote chosen by the customer can be synchronized with the Opportunity so that the products or services from that Quote are added to the Opportunity when the salesperson closes the deal.

Okay, let's cut the talking. Let's see how to create a simple Opportunity Quote:

1. **Activate Quotes**: To do this, go to **Setup | Quotes Settings** and check the checkbox next to **Enable Quotes**. Also, add the Quote Related List to Opportunity Page Layouts in the next step.

Figure 11.19: Enabling Quotes

2. Open any Opportunity, locate the Quote Related List (likely found on the right side and the last one), and press the **New Quote** button.

Figure 11.20: Creating a new Quote

3. Fill in the information about the Quote, including its name and the expiration date. In my case, the Quote will consist of the GenWatt Diesel 1000kW product and the SLA Bronze service, so I have named the first Quote accordingly.

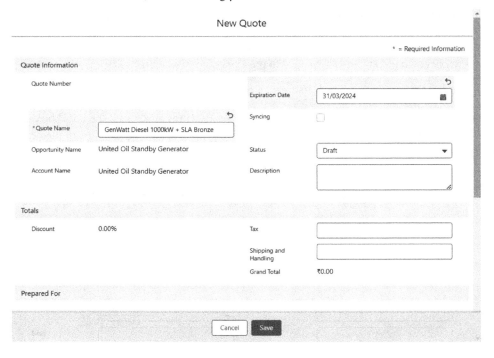

Figure 11.21: Details of the new Quote

4. After creating the Quote, add the products or services it should include. As you can see, Quotes are records that offer great flexibility. In addition to adding products to it, you can change its stages, allowing, as mentioned earlier, internal processing of the Quote or collaboration with the customer.

Figure 11.22: Adding Quote Products

5. Choose Products, click the **Next** button, and confirm the Product quantity to add them to the Quote.

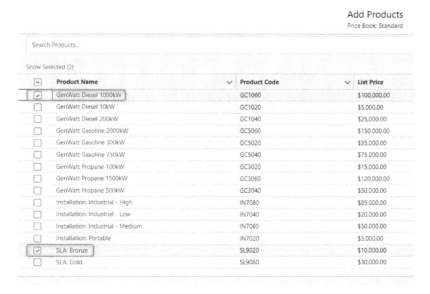

Figure 11.23: Selecting Quote Products

6. Create a second Quote on the Opportunity – this time name it `GenWatt Diesel 2000kW + SLA Gold`, and add the respective products or services to the Quote in the same way you added them to the previous Quote.

7. Ultimately, on the Opportunity, you should see two Quotes – this makes it clear that two different offers with various products and services were prepared for this customer. Everything becomes more transparent and readable.

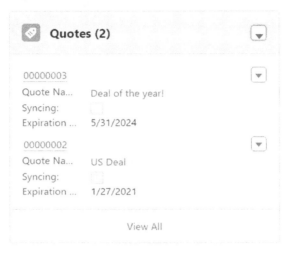

Figure 11.24: Opportunity with Quotes

Salesforce Quotes additionally allows us to do a few things that facilitate the work of the sales teams:

*   **Create a PDF**: Yes, you can generate a PDF document from your Quotes and use this document to present to the customer.

Figure 11.25: Opportunity with Quote PDF

- **Email Quote:** You can email your Quote directly from Salesforce! This is even possible directly from the PDF preview. Just use the dedicated **Save and Email Quote** action button.

- **Start sync:** The products from the Quote will be copied to Opportunity Products and synced so when you change something in the Quote Products, those changes will be applied to the Opportunity.

As you can see, Quotes are a very interesting functionality that can be useful to any sales department. They complement the standard features of Opportunities and allow for their broader utilization. They enable the implementation of more complex processes related to approvals, the generation of offers in PDF format, or sending them directly to the client from Salesforce. This means that at every stage of quoting, the user can stay within one system, which is Salesforce. Through such workflow optimization, we save time while simultaneously gathering additional valuable information.

Opportunity list view versus Kanban view

As we know, salespeople are usually very busy individuals and typically don't work on just one deal but strive to acquire as many high-quality clients as possible, for which they are usually compensated. They don't work in Salesforce on just one Lead or one Opportunity but often have dozens, and sometimes even more. To manage this, both experience and a useful system are needed, whose user interface allows for convenient work on multiple deals. Salesforce supports this through the list view functionality, a standard solution that you can also find on other Salesforce tabs, such as **Accounts** or **Contacts**. Moreover, it goes even further with the Kanban view, which can be set up on Opportunities (this view is also available on other Salesforce objects).

Let's see how the standard list view differs from the Kanban view using the example of Opportunities. In the following screenshot, I present the list view.

Figure 11.26: Opportunity list view example

As you can see, the standard list view in Opportunities doesn't differ in any way from list views encountered on other Salesforce objects. The list view allows us to change Opportunity stages in bulk, which can be a useful feature for some salespeople or certain types of sales. To do this, simply select multiple records and start typing the Opportunity stage for one of them. The following screenshot illustrates how to do this:

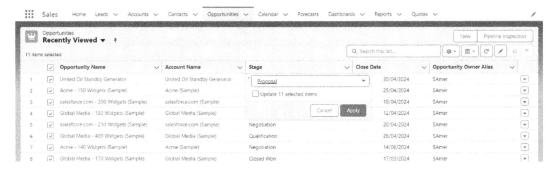

Figure 11.27: Opportunity list view mass update example

Certainly, a good solution is to create new custom list views that would show salespeople their Opportunities, for example, those they are supposed to close in the current month or quarter (using the **Close Date** field). The following screenshot shows a custom list view that presents the salesperson with their Opportunities (filtered by **Owner | My opportunities**) that are still open (filter **Closed equals False**) and are supposed to close in the current month (filter **Close Date equals This Month**). It's worth considering additional views that expand the perspective to other months, for example, creating a view that shows Opportunities closing in the current quarter. To do this, in the filter, instead of **This Month**, you can enter **This Quarter**. Simple, right? Definitely, and by the way, very helpful for the sales department. The following screenshot presents a List View limited to Opportunities that are supposed to close in the current month.

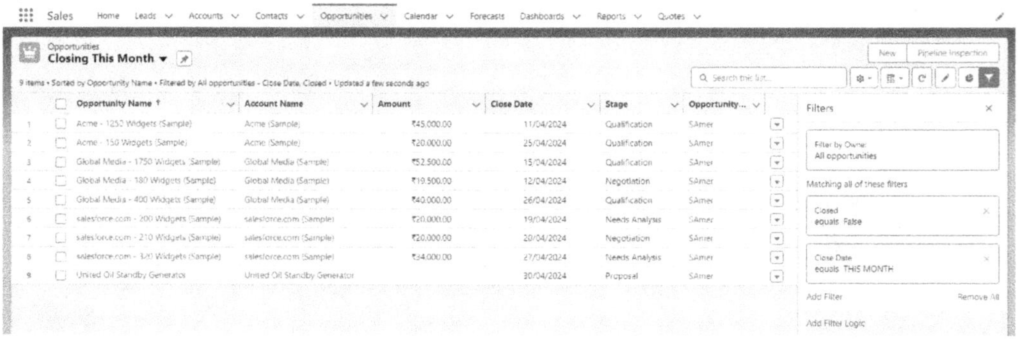

Figure 11.28: Opportunity list view custom filter example

Alright, now that we know how to use the list view on Opportunities, let's also see what benefits can come from using the Kanban view. To switch to this view, you just need to use it (perhaps a quick configuration will be necessary where we set the **Stage** field as the field to be displayed). Look at the following screenshot, which presents the Kanban view of all Opportunities I have on my Salesforce test account. Doesn't it look interesting?

Figure 11.29: Opportunity Kanban view example

Thanks to the Kanban view, we have a better visual insight into the stage of each Opportunity, and above all, we can clearly see the ratio of open Opportunities to closed ones. Of course, working in the Kanban view with a large number of records is not optimal, but Kanban can be used on any list, including one that limits the number of records, for example, to those closing in a specific month. An exceptional advantage of Kanban is the ability to change the Opportunity stage using drag-and-drop. Simply grab the tile representing the Opportunity record and drop it in another place, as shown in the following screenshot.

Figure 11.30: Opportunity Kanban drag-and-drop

Changes in Opportunity stages through drag-and-drop may appeal to some salespeople. Additionally, Kanban cards on Opportunities show alerts for overdue tasks, no open activities, and no activity in the last 30 days. If you have Salesforce Unlimited licenses, you will see even more! If you are such a lucky person (congrats!), you can activate a feature that enables, in the Opportunities list view and Kanban view, text colors and arrows to indicate amounts and close dates that changed during the last seven days. Your users will also be able to hover over an arrow to get details. How fun is that!

Kanban is a very interesting functionality that can streamline the work of the sales department. However, it's important to note that, like many features in the Salesforce world, it has its limitations. We won't list them here, as they may change by the time you're reading this book, so it's worth checking for relevant information on `https://help.salesforce.com/`.

The solutions related to list view or Kanban described in this part of the book can, of course, be applied to other objects, especially in Salesforce, such as Leads. Don't be afraid to experiment, as long as you're conducting these experiments in a test environment to avoid disrupting users who are diligently working to earn more millions for your company! Will you do this, or will your answer be more of an *"I can't promise I'll try, but I'll try to try"* (Bart Simpson, *The Simpsons*)? For sure your managers will help you with this, but let's now look at how Salesforce can help sales managers track sales numbers in the Salesforce Forecast module.

Forecasts

Now that you've learned how the sales team works with Leads, Campaigns, and Opportunities in Salesforce, it's time to see how Salesforce supports their supervisors. Forecasts is a tool or set of tools that allows sales management to track and analyze current sales progress and results. It provides valuable information related to the current state of sales, allowing you to view them by team or individual salesperson. To access Forecasts, there's a dedicated tab in the Salesforce interface and specific permissions are required. However, before you can view anything in Forecasts, it needs to be configured (you can do this in Setup). We won't go into detail on how to do this because the Forecasts functionality has many possibilities and it's difficult to describe a universal solution that will work for everyone. Just keep in mind that Forecasts are based on Opportunities, meaning that records of these objects are taken into account in Forecasts reports. You can track specific products on Opportunities, as well as setups that track Opportunity amounts. It's possible to consider Opportunity Splits, which are margins divided among sales participants. To complement this reporting, you can also set targets or Forecast quotas for each salesperson with an account in Salesforce. This allows sales management to track the current sales status versus plans and targets for each salesperson. Forecasts is indeed a very interesting feature from Salesforce, offering significant configuration possibilities. Interestingly, the interface of this functionality has recently undergone a major refresh and now looks much better than it did some time ago. Let's discuss the individual elements of this new, refreshed user interface, using the example of a salesperson, who, surprise, surprise, is me!

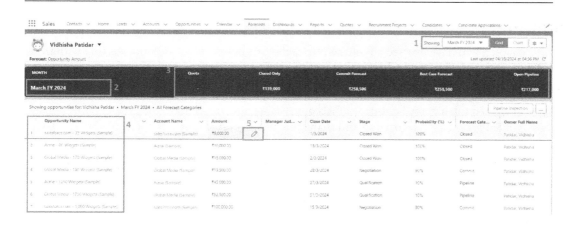

Figure 11.31: Salesforce Forecast user interface

As you can see in the preceding screenshot, Salesforce Forecasts provides a lot of valuable information. I have described the Forecasts view features as follows. Each number corresponds to the number provided in the preceding screenshot:

1. You can set the **Forecast Range** setting, which clearly indicates what is displayed on the page. If you want to change the range, simply select the **Showing** field to opt for a different range.

2. The current period is displayed in the summary view. You can change it.

3. Refreshed icons simplify page navigation, and highlighting shows the data as progress bars, clearly indicating sales numbers versus specific targets. You can also see calculations on how much is missing and the numbers related to Opportunities that can be closed soon, as they are in the advanced sales phases categorized as commit or best case (you can define them when setting up Opportunity stages). Of course, you can also hover over each highlight to click and see additional details.

4. On this list, you can see the Opportunity records related to the highlights used previously, so you can see based on which Opportunities Salesforce shows data in the Forecasts.

5. You can make changes to individual Opportunities by editing data such as Opportunity amount, close date, or stage.

Thanks to the described solutions, Forecasts is not only a static reporting tool but can also be actively used during 1:1 meetings of sales managers with members of the sales team, during which they can discuss the current sales situation of a given salesperson and make live changes if necessary. As you can see, Salesforce Forecasting is not just a simple report but a whole module that, when configured correctly (according to the requirements of your sales department) and used regularly, can provide valuable information for sales management, allowing them to have ongoing control over sales processes, the stages at which individual salespeople are, and the entire team (or teams) versus the plan imposed on them. Presented information can contribute to even better results for the sales department and, above all, allow for real planning of the next sales challenges.

Summary

In this chapter, you have learned about the intricacies of Salesforce Sales Cloud, starting with the crucial domain of Leads. You delved into the process of Lead conversion, discovering the seamless transition from potential interest to engaged prospects within the Salesforce platform.

As you navigated further, the landscape of Accounts came into view, showcasing the organizational prowess of Salesforce in categorizing and managing business Accounts. The narrative deepened, illustrating how Salesforce provides you with a holistic view of customer interactions and relationships.

Next, we talked about Activities, stressing the importance of keeping track and managing tasks and events related to customer interaction. Salesforce's Activity feature came out as a strong tool, making sure you can thoroughly monitor and proactively handle these tasks.

Then, we landed in the world of Campaigns and Campaign Members. The narrative underscored the integration of marketing efforts within Salesforce Campaigns, shedding light on the pivotal role played by Campaign Members in measuring the success of marketing campaigns and tracking customer responses.

Our further exploration then led you to the core of sales endeavors – Opportunities. You unraveled this stage of the sales cycle, which stands as a core feature, the "crème de la crème" of Salesforce Sales Cloud.

We also described the concept of Opportunity Contact Roles, enhancing your understanding of customer interactions by associating one or multiple Contacts with an Opportunity. This feature added layers to your engagement strategy, providing a nuanced approach to customer relationships.

Continuing on, you smoothly brought together Opportunity Products and Quotes, seeing how Salesforce makes it easy to add and handle products or services in opportunities. This shows that Salesforce can easily adjust to the different requirements of your sales tasks.

Finally, we discussed Salesforce Forecasting capabilities and described the Forecast interface to help you understand how sales management can track and report sales numbers.

As we conclude this chapter, Salesforce Sales Cloud has been unveiled, providing you with a comprehensive guide to navigating the intricacies of Lead management, Account organization, Opportunity exploration, and the myriad features that empower your business in its customer-centric endeavors.

Of course, we did not cover all the functionalities related to Salesforce Sales Cloud. If we were to do so, this book would need to be renamed *Salesforce Encyclopedia* and would have at least 1,000 pages. Some important topics, such as Salesforce Products, were only briefly mentioned, and others, such as Orders and Contracts, were omitted to simplify the described sales process. However, you can read more about them on the Salesforce Trailhead web page. Finally, all roads lead to Salesforce Trailhead! This is the place where the authors of this book also began their Salesforce journeys. In the next chapter, we will discuss how to harness the potential of Trailheads to learn Salesforce, along with other tips related to mastering Salesforce to obtain desired certifications, or even multiple certifications. We encourage you to read it! We strongly believe it's worth it! Remember, we are proud that you are still with us and have read another chapter because we know that in learning Salesforce, as in life, perseverance is crucial, and dealing with temporary setbacks is essential because *"life ain't about how hard you can hit, it's about how hard you can get hit and keep moving forward"* (*Rocky Balboa, movie*)

12
Salesforce Administrator Exam Preparation

The time has come, and the moment has arrived when you're nearing the end of this book, which can only mean one thing (well, it can mean many things, but one is most important): that the Salesforce Certified Administrator exam is right ahead of you. This exam is the culmination of your study journey and your hard work invested in analyzing Salesforce functionalities not only in theory but also, we hope, in practice. With this book, we wanted to encourage you not only to read about Salesforce but also to actively click through the system with us because there's no better way to learn Salesforce than through hands-on practice. As they gracefully summarize it in Star Wars, "*This is The Way*"! Personally while preparing for the Administrator exam, I made sure to click through Salesforce everywhere and everything just to see how a particular functionality I just read about looks in practice. Let's be honest, I didn't always grasp every theory related to Salesforce features from the beginning. So, I had to dive into the system (on a developer's test environment) and click around to see what would happen if I did this, and what would happen if I did something else. And there's nothing wrong with that because, as we know, sometimes instructions (such as Salesforce Help) aren't entirely clear or don't describe everything accurately.

In this chapter, we would like to share our thoughts with you on how to prepare for the Salesforce Certified Administrator exam. We'll first talk a bit about the exam format (or rather, formats, since they can vary) and provide information about scoring. Then, we'll focus on the strategies you can employ during your Salesforce study to optimally use the time you have to prepare for the exam. I'll describe one of the sources, which may not be obvious but was a game-changer for me in terms of learning and understanding the Salesforce platform. By the end of this chapter, we'll provide you with some sample test questions we've prepared and describe a few tips and tricks related to the exam itself. Let's get this party started!

We will cover the following topics in this chapter:

- Exam format and scoring
- Study strategies
- Tips and tricks for exam day

Exam format and scoring

The topic of certification was initially touched upon in *Chapter 1, Getting Started with Salesforce*, however, here we would like to delve deeper into it and focus solely on information related to the Salesforce Certified Administrator exam. After reading this book, approaching this exam should be one of your priorities. Of course, you can take the Associate exam before it, but despite its simplicity, its scope is different from the Administrator exam, and the book you are currently reading will better prepare you for the Administrator exam than the Associate exam. In this section, we will describe topics related to the format and scoring of the Salesforce Certified Administrator exam.

So, who is the Salesforce Certified Administrator exam for? According to the Salesforce exam guide, "the Salesforce Administrator credential is designed for individuals who have experience with Salesforce and continuously look for ways to assist their companies in getting even more from additional features and capabilities. The exam covers the breadth of applications, the features and functions available to an end user, and the configuration and management options available to an administrator across the Sales, Service." As you can see, Salesforce takes a comprehensive approach to the Administrator exam. This is because a significant portion of the tasks that Salesforce Administrators undertake and the tools they can use to accomplish these tasks are common across the three clouds mentioned by Salesforce. This includes tasks such as adding and activating new users, managing permissions or access, creating record views, and automation. Okay, but getting to the heart of this chapter, let me tell you more about the exam itself.

At the time of writing this book, Salesforce provides the following information about the exam:

- **Content**: 60 multiple-choice/multiple-select questions and 5 non-scored questions
- **Time allotted to complete the exam**: 105 minutes
- **Passing score**: 65%
- **Registration fee**: USD 200 plus applicable taxes as required per local law
- **Retake fee**: USD 100 plus applicable taxes as required per local law
- **Delivery options**: Proctored exam delivered onsite at a testing center or in an online proctored environment
- **References**: No hard-copy or online materials may be referenced during the exam
- **Prerequisites**: None required; course attendance highly recommended

As you can see, there is quite a bit of time available in the exam, and you don't need to score 100% (or even 70%) to pass.

To sign up for and take the exam, you only need to meet a few basic criteria: first, you have to want to do it; second, I recommend preparing for the exam; and third, you must have internet access. Simple, isn't it? The first two points probably don't need explanation, so let's focus on point number three here. To sign up for the exam, you must create an account on the Salesforce partner website called *Webassessor*, available at `https://webassessor.com/salesforce`. *Webassessor* is a platform from Kryterion, which Salesforce uses for exam registrations. The following screenshot shows the Webassessor login page. Of course, if you don't have an account there yet, you can create one using the **Create a new Webassessor™ login now** button.

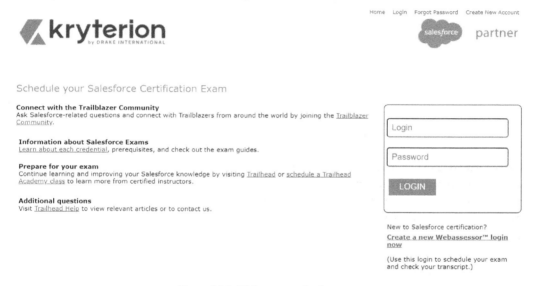

Figure 12.1: Webassessor login page

After logging in, you will see a dashboard where you can find many things, including links to Salesforce Help describing exams, their formats, and similar items. I encourage you to explore the contents of the page yourself because here we will focus on the most important functionality, which is the **Register for an Exam** tab. It is in this place that Salesforce allows us to register for exams.

Upon entering the aforementioned tab, you will see a list of all exams or exam groups conducted by Salesforce. To register for the Salesforce Certified Administrator exam, simply expand the **Administrator Exams** tab. As you can see in the following image, Salesforce offers three formats in which this exam can be conducted:

- Online Proctored
- Onsite Proctored
- Event Proctored

The three formats can be seen in the following screenshot:

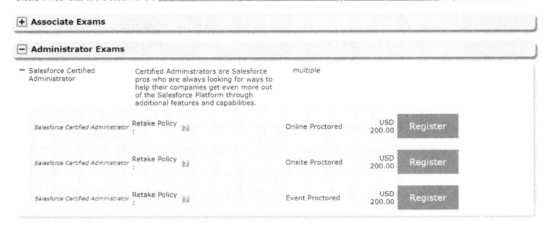

Figure 12.2: Webassessor Register for an Exam tab

The three formats shown here differ in terms of where the exam takes place and to some extent how the exam is conducted. However, they do not differ in the scope of material or types of questions that you may encounter. Besides providing a brief description of each exam format, I would like to outline the pros and cons associated with them. Nevertheless, I must say that they are subjective as I have my favorite exam format, but more on that later.

Online Proctored

The exam will take place on your computer, at your home. I won't delve too much into the technical details of this exam because they sometimes change (as was the case, for example, during the Coronavirus pandemic).

> **Tip**
> I encourage you to familiarize yourself with the current details here: `https://trailhead.salesforce.com/help?article=Online-Proctoring-Completing-Your-Exam-Remotely`

- Pros:
  - You can use your own computer, which you are familiar with
  - You can take the exam in the comfort of your own home, which can be less stressful
  - There is no need to travel anywhere to take the exam, so it's cheaper

- Cons:

  - You need to use your own computer, which you are familiar with (I know that I have mentioned this as a pro but read the end of this sentence to discover why this can be a disadvantage), but it may not be functioning properly and could malfunction during the exam.

  - You need to install dedicated Salesforce software before the exam.

  - You will be proctored via your own webcam, so you need to focus on the screen and avoid looking around too much. If you do, you may be asked to redirect your laptop camera, which can disrupt your rhythm of answering questions.

  - If there is an issue with your computer, you are responsible for it. As a result, if the exam is interrupted due to technical issues on your end, you may lose the money invested in the exam.

Onsite Proctored

The exam will not take place on your computer, not in your home, but at a Salesforce partner's facility, which is responsible for administering the exam. You will be able to choose from a list of locations that Salesforce collaborates with within your region or even city. For example, such a view appears when I try to sign up for an Onsite Proctored exam.

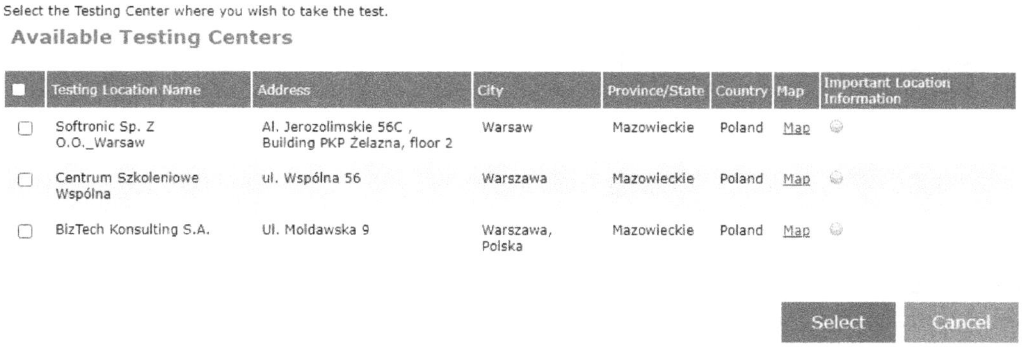

Select the Testing Center where you wish to take the test.

Available Testing Centers

☐	Testing Location Name	Address	City	Province/State	Country	Map	Important Location Information
☐	Softronic Sp. Z O.O._Warsaw	Al. Jerozolimskie 56C , Building PKP Zelazna, floor 2	Warsaw	Mazowieckie	Poland	Map	☺
☐	Centrum Szkoleniowe Wspólna	ul. Wspólna 56	Warszawa	Mazowieckie	Poland	Map	☺
☐	BizTech Konsulting S.A.	Ul. Moldawska 9	Warszawa, Polska	Mazowieckie	Poland	Map	☺

Select   Cancel

Figure 12.3: Webassessor Testing Centers in my region

As you can see in the preceding image, in the city where I live, I have three different places where I can take Salesforce exams. Okay, but what are the pros and cons of Onsite Proctored exams?

- Pros:

  - Everything is prepared for you, so you don't need to worry about things like having the right computer or internet connection.

  - You may feel more relaxed, as the entire exam room is monitored with CCTV, reducing the chance of someone else taking the same exam simultaneously and potentially seeing your answers. This allows for more flexibility in your movements and gaze during the exam.

- You are required to surrender all belongings before entering the exam room, including your watch, keys, and anything else in your pockets. Even your glasses will be checked for hidden cameras. This level of scrutiny can provide a sense of security and relaxation during the exam, knowing that you cannot cheat.

- Cons:

  - You need to travel to take the exam, as not all cities have test centers that support Salesforce certifications. This can add stress and additional costs.

  - Testing in an external exam center can be more stressful for some individuals who prefer taking exams in their own peaceful and familiar environment.

  - There may be other individuals taking different exams at the same time. Although talking during the exam is prohibited, some people may inadvertently make noise or mutter under their breath, which can be distracting. While earplugs are provided, they may not be comfortable to wear for everyone. Additionally, there may be noise from adjacent rooms, such as from other training sessions (it happened to me once). In summary, you may not be alone during the exam, and there may be increased noise levels, affecting your concentration.

- **Event Proctored** – Sometimes, Salesforce allows individuals to take exams at events they host. This is a unique opportunity to take the exam outside of your home, but often this possibility is limited to dedicated invitations. I won't describe the pros and cons here, as this form of exam is essentially the same as the Onsite Proctored exam, so they will be consistent with that format.

I promised to tell you which option I always chose. Well, my dear Salesforce enthusiasts, I took all my exams in the Onsite Proctored format. I choose this option because, in the city where I live, there are examination centers that allow taking Salesforce exams, and even more so because, during the exam, I prefer someone to take care of the computer and ensure the proper functioning of the exam platform for me. At the beginning, I told you that this is my subjective assessment and preference, so I understand that yours may be different because I know many people who always choose to take all their exams at home because they prefer it, and that's it! Fortunately, Salesforce allows for that choice, so you have the option to choose either way, and that's great. Finally, in the discussion about how to sign up for the exam, I want to inform you that the exam can be rescheduled. I have done this before when I felt that I had not dedicated enough time to study, which resulted in average scores on practice tests, or when something else came up on that date. However, remember that to reschedule a Salesforce exam, you must do so 72 hours before your exam.

TIP

Now that you know about the exam formats and where to sign up for them, I'd like to give you a hint that could enhance your comfort during the test. You can extend your exam time! Yes, you read that correctly! **Extended Time** (**ET**) exams are tailored for individuals who are not native English speakers or those with learning disabilities. These exams provide an additional 30 minutes for the test duration compared to standard certification exams. To avail of this option, you'll need to create a Case with Trailblazer Help.

Hope you managed to choose a convenient exam format for yourself, thanks to our tips! Now let's take a closer look at what this exam entails in detail. What is the scope and scoring like? However, remember that these things may change, so I want to encourage you to always verify such details directly on the official Salesforce websites before your exam, and even before you start preparing for it. The scope of the exam may be supplemented with new content continuously after the release of new functionalities. At the moment, a website worth visiting to learn the details of the Salesforce Certified Administrator exam is `https://trailhead.salesforce.com/en/credentials/administrator`.

When we were writing this book, the exam scope, known as the Salesforce Exam Outline, looked as follows:

- Configuration and Setup: 20%
- Object Manager and Lightning App Builder: 20%
- Sales and Marketing Applications: 12%
- Service and Support Applications: 11%
- Productivity and Collaboration: 7%
- Data and Analytics Management: 14%
- Workflow/Process Automation: 16%

As you can see from this list, Salesforce has assigned percentage weights to each topic, indicating their significance in the exam. These percentages translate into the number of questions in each section and, consequently, your score. For example, topics related to productivity and collaboration have the lowest weight, which means that there will theoretically be fewer questions about Activities, Chatter, or the mobile app in this exam, although this doesn't imply they won't be present at all. However, these weights show you where to focus most of your attention in your studies, especially when you have limited time.

Lastly, in this part of the chapter, I want to mention one very controversial issue that can get you a red card from Salesforce. I'm talking about the use of so-called dumps. Dumps are real questions from Salesforce exams that someone has obtained in a clever but unethical way. As you know, everything can be found on the internet, so you may encounter original Salesforce questions, and some even sell them. I strongly advise you never to use such assistance, firstly, because it's cheating, and secondly, because you won't learn anything, and your lack of knowledge will quickly become apparent in a job interview or on a project. If these arguments haven't convinced you not to use dumps, I must add that if Salesforce somehow finds out that you've used such questions, you may receive a lifelong ban and lose all your certificates (even those you didn't obtain through dumps), and you won't be able to certify with Salesforce. So, say no to dumps, say yes to knowledge. Let the Trailheads be your way! Let's see in the next section what study strategies you can take to increase your knowledge and learn fast and optimally.

Study strategies

Study strategies for Salesforce don't differ much from strategies you might adopt when learning any other IT solution, in my opinion. To become proficient with a tool, you simply need to work with it extensively! Simple, right? But what if you want to learn Salesforce but don't use it regularly at work? The answer to this question is quite straightforward. You should change your job to one where you use Salesforce! Of course, this is somewhat of a joke because without knowing Salesforce, it's challenging to land a job related to the platform unless it's as a user. So, what should you do? There's nothing groundbreaking to say here – you just need to find time to click around Salesforce on your test environment (Development or Trailhead). When I was learning Salesforce, I was fortunate enough to have a company I previously worked for decide to implement it, and I participated in the implementation before eventually administering the tool. However, our implementation was relatively simple. Therefore, to explore other aspects of Salesforce, I learned in the same way I encourage you to: by clicking around extensively on my test environment. However, I did so methodically rather than chaotically, finding some structure or scenario to guide my approach. This could involve using materials or courses on Salesforce Trailhead, video courses on platforms such as Udemy, or materials from Salesforce blogs or YouTube.

As for practical application, when learning Salesforce, I initially relied heavily on video courses. However, I wasn't merely a passive viewer. Instead, I frequently paused the videos to try out the functionality presented directly in Salesforce, navigating through the same steps as shown in the course. This allowed me to independently understand what was happening in Salesforce and why, reinforcing my understanding. Additionally, I believe that typing things on the screen helps reinforce the terminology and naming conventions of Salesforce visually. Many of us are visual learners, and without this approach, I would never have remembered certain specific names of Salesforce features. Another challenge is that Salesforce often changes the names of its features. But there's nothing we can do about that. Therefore, my advice is not to be passive during learning, regardless of the source.

But can you be passive while completing tasks on Trailhead? Unfortunately, in my opinion, yes. Salesforce Trailhead tasks are usually explained very explicitly, leaving little room for interpretation. This can sometimes lead to mindlessly clicking through Trailhead without truly understanding or remembering much. I recommend occasionally pausing, and consulting other sources including YouTube videos to gain a deeper understanding of a particular functionality, especially if it's highly weighted on the Salesforce Certified Administrator exam.

Of course, you can use Salesforce Trailhead when preparing for the exam and Salesforce even proposes official Trailhead modules that can help you to prepare for the exam (we will cover them in this chapter). The only thing we do not recommend is to do the Salesforce Superbadges (special types of Trailhead exercises) as they are rather complex and based on specific scenarios. They are also time consuming and focus on some details too much to the detriment of others.

A common question for those starting their journey toward Salesforce Certified Administrator certification is *How long will it take to study?* Unfortunately, there's no one-size-fits-all answer. Everyone learns differently, has varying amounts of time to dedicate to study, and has different prior professional experiences. However, from my experience working with people preparing for the exam, I know that it's possible to do it in three months, dedicating a few hours per week to study. As I mentioned earlier, much depends on your background. Perhaps you've been or are already a Salesforce user, so certain topics will be easier for you to understand than someone encountering the platform for the first time. Before taking an exam, I recommend taking practice test questions to assess your knowledge. If you consistently achieve scores around 90% on these practice tests, you may be ready for the real exam. This approach allows you to identify any areas where you may need further study and ensures you feel confident and prepared on exam day.

If you have limited time to study but want to learn the platform quickly, I recommend using video courses rather than text-based ones. You can watch and listen to videos anytime and anywhere, even while doing other tasks. I know, I know, I said a few sentences ago that this isn't recommended or optimal, but sometimes it's necessary, and I understand that. What is important, the video courses should directly reference the official exam outline.

However, if you have more time to dedicate to learning, focus on real-life scenarios or practical uses of the Salesforce platform. I'll write more about this later in this chapter.

Remember also that Salesforce is primarily a business platform, so start by learning the business functionalities of Salesforce before diving into configuration in Setup. Before you start configuring the system, you must understand how users work on it.

Another strategy that works for some people is to schedule the exam a few months in advance, which self-motivates them to prepare for the upcoming exam within a set timeframe. However, this may not work for everyone, as it can cause additional stress. Remember, though, that exams can be rescheduled. I've done it myself.

Let's summarize the study strategies:

- Plan how much time you will dedicate to studying and try to stick to that plan.
- Remember to refer to the exam outline described on the official Salesforce page dedicated to the Salesforce Certified Administrator exam.
- If you have limited time, opt for video courses but, if possible, try to mix resources (we will talk about them more in this chapter). The resources you choose should reference the official exam outline.
- Use Salesforce Trailhead but omit the Superbadges as they are more complex and based on specific scenarios. Focus on the recommended Salesforce Trailhead courses.
- Consider booking the exam in advance as it may motivate you to study.
- Explore the test environment and see what each function does in practice by experimenting with different settings.
- Be an active learner by clicking on the features you see in the course.

- Start by learning the business functionalities of Salesforce before diving into a configuration in Setup.

- Follow a specific scenario or course, such as a video course, to organize your exploration of the system.

- If you have more time, focus on gaining a deeper understanding of the platform through real-life scenarios.

- Test your knowledge before the actual exam. We recommend achieving a score of around 85% (preferably 90%) on mock tests before taking the real exam.

You now have the Salesforce learning strategy under your belt, so let's next explore the sources you can use to prepare for the Salesforce Certified Administrator certification. I will describe a few key sources that I have personally used and still use when learning about Salesforce. Our country's president once said that he learns all the time, from everyone and everywhere. Learning Salesforce should be similar. The following sections discuss the most important learning sources, which I've divided into Salesforce resources (official materials from Salesforce) and non-Salesforce resources (materials created by Salesforce experts).

Salesforce resources

In this section, we will look into useful official sources and materials from the CRM god, which is of course Salesforce. Some of them you probably already know, but perhaps I will provide a different perspective on them for you.

Trailhead

You probably already know about Trailhead as we have mentioned Salesforce Trailhead many times in this book as well. When studying Salesforce for the Salesforce Certified Administrator certificate, it's worth focusing on the Trailhead modules recommended by Salesforce itself at `https://trailhead.salesforce.com/en/credentials/administrator`. At the moment, Salesforce recommends the following trails:

- **Prepare for Your Salesforce Administrator Credential**: This is a Trailhead Trailmix, which is a collection of several modules merged into one, covering the scope required for candidates aiming for the Salesforce Administrator role.

- **Study for the Administrator Certification Exam**: A short and simple trail. It's worth doing it at the beginning, but it's not sufficient on its own. You can learn very basic stuff about Salesforce configuration and features here.

- **Interactive Practice Test for the Salesforce Certified Administrator Exam**: Currently, it consists of 30 free test questions from Salesforce, similar to those that will appear on the exam.

Additionally, it's worth looking for other Salesforce Trailmixes created by Trailhead users. Yes, you read that correctly; anyone can create their own Trailmix consisting of various Trailhead modules and share it with others. I have used such Trailmixes myself. One example of such a Trailmix is created by Mike Wheeler, a well-known figure in the Salesforce community for creating certification preparation courses. Mike has created his own Salesforce Trailhead Trailmix called "*Mike Wheeler Administrator Certification Trailmix*," which you can find by searching within the Salesforce Trailhead platform after logging in or just by googling it. The author of this text has also created his own Trailmix focusing on the basic functionalities of Salesforce Sales Cloud. The links to both Trailmixes you will find in the *Further reading* section of this chapter.

Additional reference

In summary, when it comes to Trailhead, I recommend sticking to the official Trailhead modules recommended by Salesforce and expanding them as needed during your free time with other Trailmixes, such as the ones mentioned by me.

However, if you have even more free time for learning, I'll now recommend something that was a game-changer for me in terms of learning and understanding Salesforce. Keep reading to find out more!

Salesforce Certificate Days

In this section, I also want to share a tip on how you can save money and get a discount on taking the Salesforce exam. This is not any illegal hack, and we won't be hacking into the CEO's Salesforce account to get the needed codes. Salesforce occasionally hosts free training events called Salesforce Certificate Days, where if you sign up and participate, you'll be rewarded with a discount code for a selected Salesforce exam. These trainings cover various topics related to Salesforce exams, so if you participate in a training related to the Salesforce Certified Administrator exam, you'll get a discount for that exam. Salesforce Certificate Days are recurring events, so I recommend keeping an eye on Salesforce-related events on social media or visiting the dedicated page: `https://trailhead.salesforce.com/credentials/cert-days`. It's worth it because the codes you receive can discount the exam by up to $100, which is half the price. Michał, who is my friend and currently a Salesforce Consultant, whom I have been supporting in his Salesforce learning journey, recently told me that the Salesforce Certificate Days made a huge impact on his learning growth, understanding of Salesforce features, and exam preparation. Michał, who was a customer support manager previously but wanted to change his work profile, is now a successful Salesforce expert working for one of the Salesforce consulting Partners. Michał *could be your inspiration* to try Salesforce Certificate Days.

Salesforce Trailblazer Community

If you're not familiar with the Salesforce Trailblazer Community yet, I recommend catching up quickly because, as I've already mentioned, it was a game-changer for me when it came to learning Salesforce. To visit the Salesforce Trailblazer Community, go to `https://trailhead.salesforce.com/trailblazer-community` and select the topic that interests you at the moment. The following screenshot shows what a question on the Trailblazer Community forum looks like:

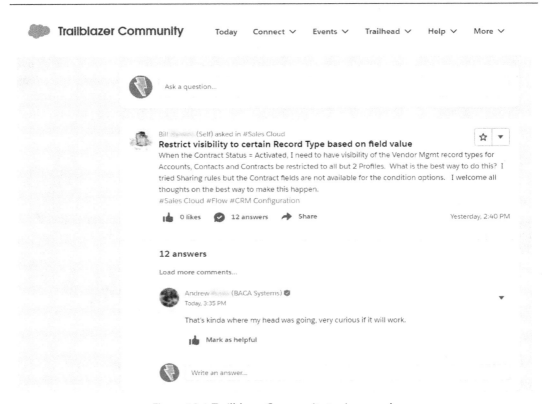

Figure 12.4: Trailblazer Community topic example

The Salesforce Trailblazer Community is a place where users ask questions and other users can respond to them! How to use it in your Salesforce learning? It's very simple! I just tried to be a person who analyzes the problems posted there by other users and provides them with answers. I wasn't always the first, and my answer wasn't always ultimately chosen by the author as correct (it was correct, but not the first correct one), but thanks to analyzing real problems or challenges faced by Salesforce users, I delved deeply into Salesforce setup. I got to know the pains of Salesforce admins, their configurations, and most importantly, I learned how they deal with creating, configuring, or sometimes even working around certain things in Salesforce. This is invaluable knowledge, especially if you don't deal with Salesforce daily and are only learning it through Trailheads. Trailheads are great, and I don't want you to misunderstand me, but Trailheads present a somewhat ideal world, an ideal setup for an ideal user. In practice, it's often a bit different, and that's what you'll discover by visiting the Trailblazer Community. Suddenly, you'll find that not every Salesforce instance is configured the same way, and even standard Objects are used in very different ways. Personally, what actively participating in the Trailblazer Community gave me was an expansion of horizons and knowledge of Salesforce setup in various ways. My previous Salesforce instance was very simple and practically unchanged out of the box. With Trailblazer Community, I could look at other setups, use other Salesforce features, and explore the challenges they entail. For example, that's where I learned the most about using Salesforce Flows, acquiring knowledge that I still use to this day. Additionally, I dealt with problems from companies in different industries, which also expanded my Salesforce skills.

> **TIP**
>
> Don't worry if your answers on the Trailblazer Community, even if correct, are not chosen as the first or marked as correct. Many people respond to questions, and firstly, I found myself providing answers, but it was my answer that was correct. I also found myself giving the correct answer, but it was not my answer that was marked as correct. Sometimes, the person asking the question forgets to mark the correct answer altogether. Remember that while such marking is very nice and certainly contributes to a dopamine boost (although maybe not as much as likes on Instagram), the most important thing is that through all of this, you are learning something new and getting to know Salesforce even better.

In the following screenshot, you can see one of the answers that I provided to a Salesforce developer from a certain company. It was helpful as he marked it as correct.

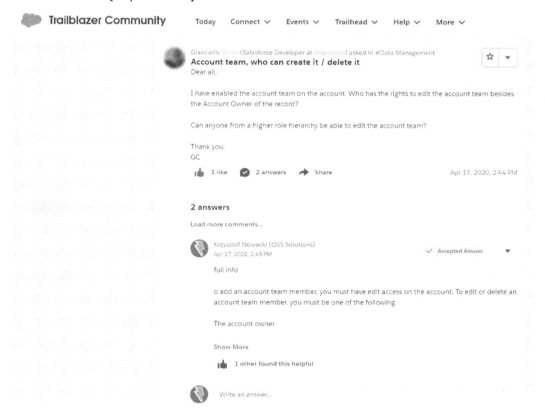

Figure 12.5: Trailhead Community Answer

I'll tell you what I really like! Salesforce is creating a community similar to the Trailhead modules. Salesforce has turned everything into a gamified form, where, as I mentioned, correct answers are marked, and ultimately, data on how many times you've helped and your answers have been accepted are displayed on your Trailblazer Profile, which looks similar to my profile in the following screenshot. It's awesome!

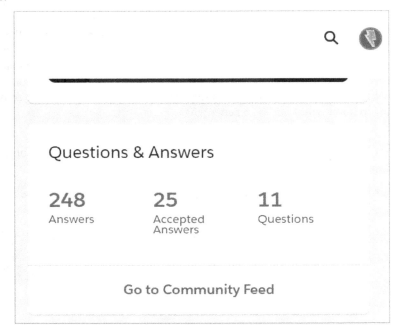

Figure 12.6: Trailblazer Community score

If you have more time available, especially if you're not currently using Salesforce in your professional work, I highly recommend checking out the Salesforce Trailhead Community. You won't regret it!

Salesforce official exam test questions

Yes, you can buy official Salesforce test questions! Where and how can you do this? You can find them on the same page where you register for the exam, that is, on the *Webassessor* page. Please see the following screenshot to see how this can be done on the Webassessor page:

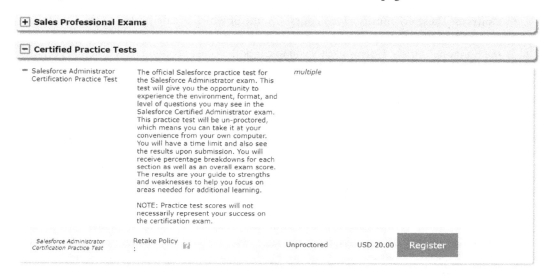

Figure 12.7: Webassessor Administrator Practice Test

As you can see, the **Salesforce Administrator Certification Practice Test** is unprocured and the current price is 20 USD. To buy access to the test just click the **Register** button.

Non-Salesforce resources

On the internet, you can find many valuable sources that can assist you in studying for various Salesforce exams, including the Salesforce Certified Administrator exam. However, remember that Salesforce recommends using their official sources and materials as per their policy, which is great, at least in theory. In practice, though, it may not be enough, and you may want to delve into a topic using materials from another source. There's nothing wrong with that as long as these are not the *dumps* mentioned previously in this chapter.

From the many places you can visit online to learn more about Salesforce, read about Salesforce features, and find well-organized legal (i.e., non-dump) study materials for various exams, I can recommend a few:

- **Udemy**: This is a learning platform where you can find courses and tests related to Salesforce:

  - **Courses**: These are mostly video courses, some of which are dedicated to the Salesforce Certified Administrator exam. There are plenty of such courses available. How do you choose the right one? Firstly, Udemy allows you to check the course content (it is described on the page of each course) and listen to a part of the video course before purchasing it, which allows you to assess whether the content provided will be suitable for you and whether it will be presented in a way that suits you. Furthermore, each course can be rated, so you can see how other participants rated a particular course and make decisions based on those ratings. Ultimately, Udemy allows for a refund if it turns out that the course does not meet your requirements. Of course, there are some exceptions associated with this, but this option is generally available. Many people from the Salesforce community use the courses available on Udemy, which undoubtedly have the advantage of being video courses where someone guides you practically through the topics related to the exam.

  - **Tests**: Similar to courses, Salesforce test questions can also be found on Udemy. It's worth checking that they have good ratings and reputable authors. I am the author of several sets of Salesforce test questions on Udemy. At this moment, my test questions cover certifications such as Salesforce Certified Associate, Salesforce Certified Administrator, and Salesforce Certified App Builder. Of course, I am not alone on Udemy, and I was not the first, so if you want to check test questions from other authors, search for `Salesforce Admin Test Questions` in the Udemy search engine, and you will find what you are looking for. However, remember that these questions do not come from an official source, so they may differ from the questions on the exam. And that's a good thing because otherwise, they would be dumps, which are prohibited. The questions that I and other creators on Udemy create are not identical to the questions you will have on the exam, but they cover the same topics, so they allow you to truly test yourself before the real exam. If you prefer to stick to official sources, remember that I described how to buy official test questions from Salesforce. However, if the price is a barrier for you or if you want to test yourself even more thoroughly, I encourage you to check out the Udemy platform.

- **Focus on Force (FoF)**: FoF is a dedicated platform for learning Salesforce. It focuses solely on Salesforce. Just like on Udemy, you can find courses preparing you directly for the chosen certification. What distinguishes FoF from Udemy is that the main materials found there are not video materials. This means that with FoF, you will have to read much more, whereas with Udemy, you can watch and listen to videos. However, FoF has a different advantage: its materials are usually more detailed than courses on Udemy, and the test questions on FoF are usually difficult and structured similarly to exam questions.

- **YouTube**: I won't recommend any specific Salesforce courses on YouTube here because I use this source in a slightly different way than courses. I use YouTube selectively when, for example, I want to better understand a specific functionality. In such cases, YouTube works very well. So, if, for example, you confuse Salesforce Profiles with Permission Sets and Permission Set Groups, you will surely find someone on YouTube who can explain it efficiently.

- **Salesforce blogs**: There are many blogs dedicated to Salesforce on the internet. It's impossible to list them all, so I'll focus only on a few that I have used most frequently. These mentioned blogs can serve as a source of information about Salesforce software itself, as they often describe practical applications of various Salesforce features or present the latest updates introduced by Salesforce. These blogs sometimes also contain test questions or teach how to prepare for exams:

  - `https://admin.salesforce.com/salesforce-admin-podcast` is the official Salesforce Administrator podcast, offering interviews, automations, low-code, and other topics related to the Salesforce Administrator role

  - `https://www.salesforceben.com/` is a well-known Salesforce blog created by the Salesforce MVP, offering tips and tricks, articles, and interviews

  - `https://automationchampion.com/` addresses topics related to automation in Salesforce

  - `https://unofficialsf.com/` is a blog containing podcasts related to Salesforce and is full of useful resources about Salesforce automation and custom features, including links to unofficial apps that you can install

As you can see, we have listed quite a few sources that may be helpful to you during your learning process. In practice, of course, you won't be using all of them, and certainly not all at once. It's best to choose one or two sources, for example, video courses complemented by Trailheads, or similar combinations.

TIP

As we know, we often learn in our free time, for example, while traveling. One thing I like about courses on Udemy, for example, is that I can save them for offline use. When I'm traveling, be it abroad, on a plane, or waiting in an airport, I can occupy my free time by watching videos about Salesforce.

If you prefer reading instead of watching videos, then check out such courses and also enrich them with practice by clicking around Salesforce. That's the most important part! Happy learning. Once you're ready, we invite you to test your knowledge in the next part of this chapter with a few sample test questions. Let's see whether you find them difficult or easy.

Practice questions and mock exams

In this section, we would like to show you what exam questions may look like and at the same time, test your knowledge a bit. This way, you'll be able to assess your level of preparedness for the real exam. Of course, the questions we'll provide are not identical to the ones you'll encounter on the exam, but they do cover the same topics outlined in the official exam outline provided by Salesforce, so they are very similar to the real questions. At the end, we'll provide the answers so you can check whether you answered correctly. So, *"Are you ready to rumble?"* Round 1! Let's go!

1. The customer care manager of Cosmic Enterprises would like to increase the visibility of open Cases to help the support team work more efficiently. The users should be able to assign themselves and change the **Case status** of Cases related to the **Equipment** record type directly from the list of Open Cases.

    What could the Salesforce Administrator propose to support this request?

    Choose three answers:

    A. A List View showing Open Cases could be created and used by the customer care agents

    B. A Salesforce report showing Open Cases could be created and used by the customer care agents

    C. To be able to inline edit the Cases in the list, the proper record type needs to be used in the filters or List View

    D. The List View can be placed on the Salesforce Home tab to support quick access to the list

2. The Salesforce Administrator has received unexpected information about a person leaving the company. What should the Salesforce Administrator do primarily to safely switch off the user's access, along with handling all potential dependencies?

    Choose one answer:

    A. Deactivate the user

    B. Change the user's password

    C. Freeze the user

    D. Completely delete the user account

3. The system administrator has received a request to activate 30 new Salesforce users. What should the Salesforce Administrator do to accomplish this task in the most optimal and fastest way?

    Choose three answers:

    A. Check that the company has all the needed SF licenses to be able to activate those users

    B. Be sure that you know which Profiles should be assigned to the users as this is a mandatory step

    C. Create all users one by one using the Salesforce user interface to be sure that all data is inserted properly

    D. Use Data Loader to create all of the requested users

4. The company Universal Containers keeps its records private, so only the record owners and their supervisors have access to this data.

    Keeping in mind that the company is using a private sharing model, how should the system administrator approach this request?

    Choose one answer:

    A. Create a proper Role structure

    B. Use Permission Sets to grant access to all the records

    C. Update the Manager field on the user profile

    D. Create a custom manager master-detail field on the user profile

5. How can the Salesforce Administrator set up custom accounting periods in Salesforce for financial and tax reporting purposes for the company?

    Choose one answer:

    A. Set Fiscal Year in Salesforce setup

    B. Set Financial Year in Salesforce setup

    C. Set Reporting Year in Salesforce setup

    D. Set Calendar Year in Salesforce setup

6. What is Salesforce OWD responsible for?

    Choose two answers:

    A. Configures the Open Wide Dynamic to control Salesforce Outlook integration

    B. Configures the default object accesses separately for different objects

    C. Configures the On-Demand Workforce to control access to the Salesforce Workforce object

    D. Configures the baseline access for your internal users for your records

7. Universal Containers would like to provide its Sales Representatives with the ability to track activities performed towards prospects or customers. Sales reps should be able to log and schedule calls and calendar meetings with prospects or customers. What considerations need to be taken into account while fulfilling this request?

    Choose two answers:

    A. The Salesforce Campaign Member object could be used to support this requirement

    B. Salesforce Tasks and Events can be used but the users need to know that their comments stored on the Task and Event records will be always editable to all the users with at least edit access to Tasks and Events parent records

       C.   Salesforce Activities (Events or Tasks) can be used, and can be configured to limit the editability of the created records

       D.   Events marked as private via the Private checkbox are accessible only by the user assigned to the event

8.   When using the Salesforce Data Import Wizard feature for data import, which of the following statements are true?

    Choose two answers:

       A.   You can import up to 50,000 records at a time

       B.   You can't import Campaigns via Data Import Wizard

       C.   You can import up to one million records at a time

       D.   You can't import Leads via Data Import Wizard

9.   You have been asked to add a few already created fields on the out-of-the-box Account record page. How will you perform this task?

    Choose one answer:

       A.   Edit the Lightning page and use Dynamic Forms while adding the fields to the user interface

       B.   Edit the Visualforce page and use the classic page layout while adding the fields to the user interface

       C.   Edit the Lightning page and use Dynamic Actions while adding the fields to the user interface

       D.   Edit the Aura page and use code change while adding the fields to the user interface

10.   You have been given the responsibility to automate particular actions that will result in the removal of specific Salesforce records. The marketing manager intends to perform daily deletions of Leads who have requested their data to be removed. Which Salesforce low-code feature can support this request?

    Choose one answer:

       A.   Approval Process

       B.   Apex Trigger

       C.   Apex Class

       D.   Flow

Answers:

1. A, C, D. Explanation: `https://help.salesforce.com/s/articleView?id=sf.lex_find_list_views.htm&type=5`

2. C. Explanation: `https://help.salesforce.com/s/articleView?id=sf.users_freeze.htm&language=en_US&type=5`

3. A, B, C. Explanation: `https://help.salesforce.com/s/articleView?id=sf.adding_new_users.htm&type=5`

    `https://help.salesforce.com/s/articleView?id=000327115&type=1`

    The fastest way to add a large number of users is to create them with Data Loader rather than creating them one by one in the Salesforce user interface.

4. A. Explanation: `https://help.salesforce.com/s/articleView?id=sf.admin_roles.htm&type=5`

    `https://trailhead.salesforce.com/content/learn/modules/data_security/data_security_roles`

5. A. Explanation: `https://help.salesforce.com/s/articleView?id=000325037&type=1`

6. B, D. Explanation: `https://help.salesforce.com/s/articleView?id=sf.admin_sharing.htm&type=5`

    Internal organization-wide sharing defaults set the baseline access for your internal users for your records. You can set the defaults separately for different objects.

7. C, D. Explanation: `https://help.salesforce.com/s/articleView?id=sf.security_sharing_considerations.htm&type=5`

8. A, B. Explanation: `https://help.salesforce.com/s/articleView?id=sf.data_import_wizard.htm&type=5`

9. A. Explanation: `https://help.salesforce.com/s/articleView?id=sf.dynamic_forms_considerations.htm&type=5`

10. D. Explanation: Apex could be used to perform this task, but the requirement is to use a low-code Salesforce solution. Flow is a low-code Salesforce feature, whereas an Apex solution would involve custom code development. Approval Process is not used to delete records.

Okay! We're done! So, what's your score? How many correct answers did you get? All of them, I hope! Even if not, remember, that it's just a test and not the real exam, so there's time to learn and improve in areas where you feel your knowledge is not yet at the level you'd like. That's why we test our knowledge, to see where we stand and make necessary corrections if needed. Above all, regardless of the result of this test, believe in yourself and keep learning. Everyone has their own pace, their way of learning, and other things on their plate besides learning Salesforce. Remember, *"Possess the right thinking.*

Only then can one receive the gifts of strength, knowledge, and peace" (Sprinter, Teenage Mutant Ninja Turtles – The Movie). Now that you have tested your knowledge, we will share tips and tricks with you regarding the day you will be taking your exam in the next section of this chapter.

Tips and tricks for exam day

In this section, we would like to present you with a few tips and tricks that may come in handy on the day of the test, while you're solving it, and even the moments just after! I've divided them into tips for just before the exam, during the exam, and after the exam. I encourage you to read these tips and tricks.

Before the exam

The following are some things to keep in mind before the exam:

- If you've opted for an online exam at home, check and prepare your computer in advance. Remember that Salesforce requires the installation of a dedicated application, and you must have a functional webcam. You should check all of this in advance.

- If you're taking the exam at an exam center, arrive early. Keep in mind that the process of verifying your data (i.e., checking if it's really you) takes a bit of time.

- Leave your gadgets in the locker at the exam center. Personal items, as well as your phone or watch, won't be needed during the exam, so leave them in the locker. You'll lock it up yourself and you'll be given a key. If you're unsure about their safety, simply don't bring valuable items with you. However, remember that you must bring two photo identification documents and your exam registration code (you'll receive this via email).

During the exam

Keep the following in mind during the exam:

- Mark questions you're unsure about. The exam tool allows you to mark such questions and return to them at the end of the exam. I often use this solution and review my answers at the end. Have I ever changed a previously marked answer? Definitely yes!

- Watch out for tricky questions and look for keywords. There aren't that many very tricky questions, but sometimes the devil is in the details, and one word or phrase used in both the question and the possible answers can determine the correct response.

- If you don't know something, skip the question until later. Don't spend too much time on one question; remember that you have a limited amount of time.

- Try not to stress about it if, for example, a few questions in a row were difficult for you. I've encountered such situations at least twice. Firstly, it may turn out that the remaining questions will be easy for you, and secondly, you don't need to have 100% correct answers. Remember that you don't have to be perfect to pass the exam; it's enough to correctly answer 65% of the questions.

- Try to remember the topics of the questions that were difficult for you. Of course, I hope that you will pass the exam on the first attempt, but if not, you'll be able to return to this book and the Trailheads and practice the topics even harder that caused you trouble.

- Keep track of the number of questions you're sure you've answered correctly. You can have a piece of paper and a pen during the exam. They'll be waiting for you at the exam center. I've encountered this strategy in one of the Salesforce groups online. I've never used it myself because I think it wastes valuable time, but some people may like this method because if the number of correct answers keeps increasing, it may calm you down and convince you that you'll ultimately pass the exam. The problem arises when the number of correct answers doesn't increase, which may stress you out even more and affect your concentration when answering the next questions.

After the exam

After the exam, keep the following in mind:

- **You didn't pass! What now?** Let's start with a more pessimistic scenario. If you didn't pass, don't worry! (I know, easier said than done.) You'll receive the result along with detailed information about the percentage you scored in each section of the exam. Analyze these numbers and focus on studying the topics where you performed weakest again. Remember that, as mentioned in this section, these topics have different weights, and some are scored higher than others, so focus on your weaknesses with the questions with the highest weight first.

- **You passed! What's next?** And now for the fully optimistic scenario! If you passed the exam, celebrate it! The Salesforce Certified Administrator certificate is a valuable achievement! You can be proud and share your accomplishments more widely, for example, through a post on social media (such as LinkedIn). You can also update your LinkedIn profile by adding information about your new certification.

- If you've passed the Administrator exam, start studying for the App Builder certificate. If you've passed the App Builder exam, start studying for the Sales Cloud certificate. Why? Because these exams are related to each other, even to the extent that some of the topics overlap. It'll be easier for you to take the exams mentioned here if you go straight ahead before the knowledge you've gained partially evaporates. The same goes for other Salesforce exams because similar dependencies exist, for example, between architectural exams.

I hope these tips will help you navigate through the exam smoothly, and nothing will surprise you on the day you take it! Remember that we're rooting for you! Be sure to reach out to us after your exam! We'll be happy to hear about your success! And if not (although you'll definitely succeed!), we'll gladly advise you on what else you can improve.

Summary

In this chapter, we aimed to provide you with information related to the Salesforce Certified Administrator exam. By reading this chapter, you have now become familiar with our recommended study strategies for the exam and will be able to apply them in practice. We also provided detailed insights into how to study Salesforce and which sources to use in your learning journey. Importantly, our recommendations are based on the sources and methods that we use and find effective. To boost your confidence when taking the exam and test yourself before the real deal, we have enriched this chapter with sample test questions and directed you to places where you can find more such questions. In concluding this chapter, we offered you practical advice related to the day of the exam, which will help you navigate through it smoothly and with less stress. We are confident that you will achieve a positive result! Taking the Salesforce Certified Administrator exam is a significant step toward advancing your career in Salesforce. With the right preparation and mindset, you can confidently approach the exam and demonstrate your proficiency in Salesforce administration. Remember to stay focused, manage your time effectively during the exam, and trust in your abilities. By following the strategies outlined in this chapter, you'll be well equipped to tackle the exam and achieve success. Keep practicing, stay motivated, and you'll soon join the ranks of certified Salesforce professionals. Good luck with the exam! *"May the force be with you!"*

Further reading

- Krzysztof Nowacki Sales Cloud Trailhead Trailmix:

  `https://trailhead.salesforce.com/users/knowackisfdc/trailmixes/get-started-with-salesforce-business-process-lightning`.

- Mike Wheeler Administrator Trailmix:

  `https://trailhead.salesforce.com/users/mikewheeler/trailmixes/mike-wheeler-administrator-certification-trailmix`

13

Continuing Education and Career Development

In the previous chapter, you learned about certifications and how to prepare for them. But I want to remind you of one thing: you work in IT, where everything changes from minute to minute. We, the IT people, need to stay up to date. So, in this chapter, we will discuss where to find out about upcoming updates to Salesforce, where to find people with similar interests, which industry events to attend, and which direction your career path should take. This chapter includes covers the following:

- Staying up to date with Salesforce releases
- Joining the Salesforce community
- Advanced certifications and career pathways

Staying up to date with Salesforce releases

I have often stressed that today's world is very dynamic. Just when we think our knowledge is up to date, a new update appears, and we start our learning journey anew. Every manufacturer of electronic equipment or software must improve their product; without this, it will become outdated and fall behind current standards, and its security will be compromised. Just as phone manufacturers release updates, Salesforce also has such updates. (Actually, that's not surprising: even my vacuum cleaner updates itself from time to time.) So, I do not mean to worry you, but after passing your admin certification, you still need to learn. But thanks to this, you can implement the latest solutions in your org. So, there is some give and take.

What is a release? A release is nothing more than an update to the Salesforce platform. These updates are released three times a year – in spring, summer, and winter. These updates provide new features, enhancements to existing features, fixes to existing bugs, and updates for system security and performance.

> **Tip**
>
> If you have an idea for an innovation that would be useful in your system but is hindered by Salesforce's limitations, visit `ideas.salesforce.com`. There, specialists share ideas for system improvement. You will find plenty of great ideas there. You can vote for the ones you particularly like. It looks something like the following:

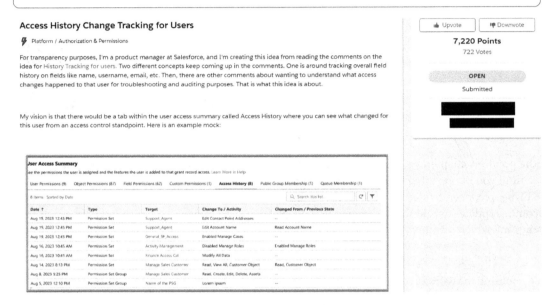

Figure 13.1: Salesforce IdeaExchange

There are a few distinctive elements related to Salesforce releases:

- **Predictability**: Salesforce does not run late; it is already waiting for you to arrive at the meeting. Seriously though, our cloud-blue producer plans the update of its system and informs users about it well in advance, giving plenty of time for admins and developers to prepare for the upcoming changes.

- **Release notes**: This is the documentation that Salesforce prepares for each system update. It contains descriptions of new features, system enhancements, or other changes to the currently available functions. This document is a source of knowledge about changes in the system. Often, it is a very fascinating read for us Salesforce freaks!

- **Sandboxes**: Before releasing changes into production, Salesforce provides customers with sandboxes implemented with the changes. This allows companies to test new functionalities without worrying about their current configuration. Moreover, they can test new features with their current system configuration.

- **Trailheads**: How great is it that Salesforce came up with trailheads – what would we do without them? And I mean that seriously, because not only did many of us start our Salesforce adventures there, but after each release, certificate holders must carefully read the maintenance trailhead and answer the questions found there correctly. This is how Salesforce encourages its users to educate themselves on new functionalities.

As you can see, Salesforce does a lot to reach its users and convey information about what will be implemented in upcoming updates. And the updates themselves are sometimes dreams come true for admins, consultants, and developers. I remember when they announced Dynamic Forms for objects other than just custom ones. I declared then that it would be my personal game changer, and I was not wrong. When designing systems for my clients, I use them on most objects. But what significance do updates from Salesforce have? There are three areas of impact:

- **Security**: Remember, we stand on the side of light, and people are trying to hack our systems every day. Well, it is not *that* bad, but fortunately, Salesforce regularly checks its security, designing new security solutions and patching gaps in the system in the process.

- **Innovation**: As I mentioned to you in the first part of this chapter, Salesforce is constantly evolving, building new functions, and improving current ones. Thanks to this, it is a leader among CRM solutions, and companies translate these new functions into even better customer relations and business conduct.

- **Compatibility**: In *Chapter 9*, you read about AppExchange, so you know that applications from external producers give even more functionalities to Salesforce. Therefore, Salesforce reciprocates by ensuring compatibility with new external applications.

> **Tip**
>
> If you want to supplement your knowledge of upcoming releases, I encourage you to check out a portal that you have probably used at least once in your life: SFBen at `https://www.salesforceben.com/`. Ben and his team discuss very well what is in any upcoming release and describe the impending novelties.

Awareness of what is coming in an upcoming update is extremely important. Without it, we could be seriously surprised when trying to create a new process in Process Builder (*Flow, I am your father*), for example, if an update introducing Flow as the main tool for automation in Salesforce came in. For Salesforce system administrators and developers, tracking new updates is a key element, allowing them to utilize the potential of the latest functionalities and maintain their systems at the highest available level of operation and security.

Preparation for upcoming releases for companies with large orgs is an extremely labor-intensive and time-consuming endeavor. During such preparations, all key solutions in the system are tested on a sandbox provided by Salesforce. Without it, many companies could lose data, time, and above all, money. Imagine that someone took all of your data about your family and friends – do you remember all of their numbers? I do not. I know a few that I would be able to recall. It would be the same in the case of not preparing your environment for an upcoming release. Imagine that Salesforce releases a new data structure, and to be able to convert to it, you need to perform a few uncomplicated actions. But you and your company do not do it, and one sunny day, the Sales department comes in and sees that the data is unreadable and most of the values from various fields have disappeared. Can you imagine that chaos? That is exactly why Salesforce releases a summary of their work and upcoming novelties.

I hope you now understand how important these release notes are before the final release. Speaking of reading, maybe you'll also write a bit and talk to people who love Salesforce just like you. In the next section, you will read about where you can find communities, events, and much more.

Joining the Salesforce community

Every passion has its community, where you will find like-minded people for whom the language you use at work does not sound like *blah, blah, blah* (shoutout to my wife, who hears me like that when I talk about my work – I will teach you Salesforce one day, honey). Just as Metallica fans have their meetups and concerts, and motorcycle enthusiasts have their events, so do we.

But let us start first with people. After all, it is us, the users, administrators, developers, architects, and many, many others who create new products and configurations and share ideas – yes, it is us who are the community. What is definitely unique in these groups is the willingness to help. During my adventure with Salesforce, I noticed that people from different international communities, when they meet, talk like old friends even though they are meeting each other for the first time. I remember my first Salesforce World Tour in Warsaw, where I met many people who I had read from on various groups or had given virtual high fives on LinkedIn. It was incredible. After constantly hearing that corporations are an eternal rat race, I saw that it was not true. These people were helpful, always wanted to help, and always did.

So, what exactly is a community? It is a collection of people, of various ages, origins, and genders, with one interest – in our case, Salesforce. There are local communities: in my case, the local community is formed across cities such as Kraków, Warsaw, Lublin, and many other Polish cities (I'll take this opportunity to greet you all!). However, the Salesforce community is not limited to a city, country, or continent. We operate worldwide – so if you have a question or can't deal with something, just search for the problem, and I guarantee you that one of the first results will be discussions in the community, where the topic has been broken down into atoms and put back together to be solved. If the problem isn't there because of a Salesforce limitation, that problem will definitely be solved.

What do I encourage you to do? Well, look for such a community in your city and your country. Why? Because it will really give you a lot. You can be an extrovert or an introvert – it does not matter, because from such a community, you can not only draw information and job offers but create friendships with people, which will be something more important than a few more dollars in your account.

For example, the group where I met my esteemed co-author – Krzysztof Nowacki – was a group of Salesforce enthusiasts founded by him. It is called Salesforce Careers Poland, and it is where we discuss topics related to Salesforce and IT, but also very human topics, related to the country, family, and many others. So, we invite you to join the group!

The second equally important community is the Trailblazer Community. I think I do not need to remind you about Trailheads. Currently, it is one of the best encyclopedias of knowledge about Salesforce. However, the Trailblazer Community consists of discussion groups created to discuss specific topics in specific places around the world.

To start your adventure with the Trailblazer Community, go to `https://trailblazercommunitygroups.com/`, which looks as follows:

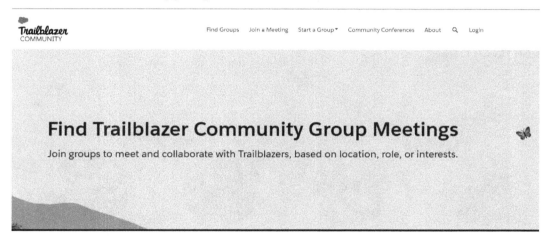

Figure 13.2: Trailblazer Community

You have several options for finding the right content for you:

- **Join a Meeting**: You often search by the city in which you would like to participate in a Salesforce event. There can be many types of events, from Dreamforce and World Tour Events to smaller SF Summits organized in smaller groups – sometimes by companies and sometimes by the community itself. These events will allow you to find different perspectives on the technologies you already know, premieres of new stuff from Salesforce, or people who want to see you and have a drink with you... lemonade! The events can vary in size and location. One of the most important is Dreamforce, which takes place in San Francisco. The event lasts ~3 days, during which you can listen to over 1,500 sessions packed with knowledge. For slightly smaller events, there's the aforementioned Salesforce World Tour, held in various cities with local partners presenting at their stands, a whole day filled with fantastic lectures and ending with social integration. Such smaller meetings are a chance to meet people in the industry and potential new clients – because it is a brilliant place for generating leads (I'm talking to you, sales department!).

The last type of meeting is Meetup events. These are meetings organized by someone from the community or Salesforce partners. They can be for 100 to even 1,000 participants. Recently, the first Polish Dreamin premiered, which took place in Wrocław, organized by Lukasz Bujlo (Salesforce MVP 2023) from Coffee & Force. The event will be held regularly, and I'm inviting you to it. It was full of brilliant lectures, Salesforce partners, and Salesforce enthusiasts. Last but not least, there are smaller events of about ~100 people organized by partners, where lectures are delivered by employees of the partner on various Salesforce topics. Ah – I have not mentioned online events yet! Those happen too. Most often, it's a day filled with online lectures, where presenters from different parts of the world have their slots to talk about Salesforce technology topics. Phew, I think that's all of them.

- **Find Groups**: These are national or city groups. Just enter your city, and you will find the group that interests you. It looks like the following, where I searched for groups in London for you.

Figure 13.3: What Find Groups looks like

- **Content**: Look for the magnifying glass icon at the top and enter the topic you are interested in, then find a group with that theme. In the following screenshot, you can see the groups that appear after typing `Nonprofit`.

Groups

Salesforce Nonprofit User Group, Adelaide, Australia
Adelaide, SA (AU)

Salesforce Nonprofit User Group, Adilabad, India
Adilabad(U), TG (IN)

Salesforce Nonprofit User Group, Amsterdam, Netherlands
Amsterdam, NH (NL)

Salesforce Nonprofit User Group, Atlanta, United States
Atlanta, GA (US)

Salesforce Nonprofit User Group, Auckland, New Zealand
Auckland, Auckland (NZ)

Figure 13.4: Thematic groups

Check out this page; it will allow you to find additional materials and people who share the same passion as you. Now, let us hop into the DeLorean from *Back to the Future* and head to the future, where you have taken all these tips and become an expert. So, now it is time for your own group. This option exists too; just click on **Start a Group** on this page, and...you might receive a message saying, *"Applications are closed at this time. Please check back on April 1st, 2069, for when the next application cycle opens."*

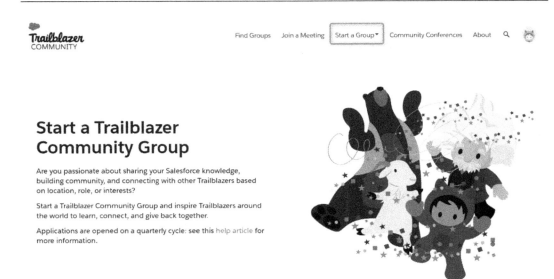

Start a Trailblazer Community Group

Are you passionate about sharing your Salesforce knowledge, building community, and connecting with other Trailblazers based on location, role, or interests?

Start a Trailblazer Community Group and inspire Trailblazers around the world to learn, connect, and give back together.

Applications are opened on a quarterly cycle: see this help article for more information.

Figure 13.5: Start Trailblazer group page

But do not worry, places will open soon! I assure you that the deadlines are not as far off as *that* message suggests. Okay, time to return to our current times. So, when you have embedded all the knowledge from all the previous chapters and have learned how to approach exams in the last chapter, it will be time to choose your career path.

Advanced certifications and career pathways

If you have already passed the Salesforce Admin certification, you are on a very good path to continue this winning streak. As they say, "*No risk, no champagne.*" Certifications are a bit like tattoos: once you get one, it is hard to stop. But what is the next certification? This question troubles many administrators. I can only advise you; I cannot decide for you. Have you ever watched *Dragon Ball*? (If not, you have a lot to catch up on.) In it, the main character and his friends reach a new levels of power called Super Saiyan, and there's Super Saiyan II and Super Saiyan III too. It is similar to Administrator certifications. Why not go for Advanced Administrator? This is the first certification that can also be useful in an administrator's near future. But there are many certifications. I will mention only those that are extremely useful and those that you could not even have dreamed of when you first learned what Salesforce was.

Advanced Administrator is a certification for experienced admins. Yes, you read that right – experienced admins. But imagine going to a recruiter with such a certification. It would be like being a 20-year-old with 30 years of experience. With this certification, you will receive a badge that looks like this:

Figure 13.6: Advanced Administrator badge

It looks awesome! By studying for this exam, you will gain knowledge about advanced security configurations, data sharing, and advanced field configurations. You will acquire secret knowledge about Flow and business process automation. You will be able to manage data, reports, analytics, and changes like an expert.

With the **Sales Cloud Consultant** certification, you will gain the skills of a consultant. Of course, a bright beam of light will not descend upon you; your soft skills will not suddenly increase by 10 levels. But you will learn to design based on one of the most popular clouds – Sales Cloud. Here is the badge for it:

Figure 13.7: Sales Cloud Consultant badge

This certification also involves data management and solution design. In addition, you will learn a lot about sales processes and the capabilities of leads and opportunities. Products and price books will no longer be foreign to you. Plus, you'll know about sales analytics and reports. What is more? Sales, sales, sales again!

With the **Service Cloud Consultant** certification, you will be a master in setting up help desks. This is another very popular direction for newly minted administrators and another good-looking badge for your collection:

Figure 13.8: Service Cloud Consultant badge

You will become familiar with case management and knowledge bases. You will be able to automate service processes using Flow to more efficiently manage cases and improve customer service quality. Setting up new contact channels will be exceptionally easy for you – chat, phone, or email – configuring them will not pose a problem for you.

Now, let us move on to the less popular exams. We will start with my favorite (which I took twice: it was the only exam I did not pass on the first try – my nemesis!).

The **Nonprofit Cloud Consultant** exam will allow you to gain knowledge about a remarkably interesting solution, which is Nonprofit Cloud – a cloud for **non-governmental organizations** (**NGOs**) and charities. The badge looks very neat:

Figure 13.9: Nonprofit Cloud Consultant badge

This exam is demanding because knowledge of the cloud alone is not enough – you also need to have an awareness of Marketing Cloud tools and Accounting Subledger. I also recommend you read a bit about NGOs and best practices in this sector. But so as not to scare you too much, remember that helping NGOs/charities is a noble goal. Sometimes, I take a few extra hours of my time to implement Nonprofit Cloud, just so I can feel like I have done something good for the world – I recommend it!

Then there is the **Education Cloud Consultant** certification – back to school! Would you like to make a system for your former university? Or maybe a school? This certification gives you a chance to show your teachers that you have achieved something in your life! (Greetings to all my teachers – see? I wrote a book!) This unique certificate has a badge quite like all the others:

Figure 13.10: Education Cloud Consultant badge

This certificate gives you knowledge about the structure of this solution. You can set the life cycles of students/pupils, adapt the platform to new educational requirements, and sometimes change someone's bad grade to a good one. I know you would do it.

Field Service Lightning Consultant is a certification for functionalities such as Field Service Lightning, which involves creating solutions for managing fieldwork. Sounds interesting, right? Here is the badge:

Figure 13.11: Field Service Consultant badge

This certificate focuses on scheduling and optimizing field tasks and managing resources and routes for the company. This time, we will play with a mobile app, as this knowledge will be tested. When reading Trailheads about FSL, always try to remember the best practices.

Platform Developer I is a certificate for those who enjoy coding. If you are someone interested in coding and have the skills for it, this certification is for you. You started with Admin, but you can comfortably move on to something more advanced, earning yourself a cool badge:

Figure 13.12: Platform Developer I badge

During your learning, you will not only understand the data model and navigate through the platform, but your knowledge of writing Apex classes, triggers, and unit tests will also be tested. Does Visualforce ring a bell? If yes, that's good because knowledge of building components and integrating them with Apex will be required. Knowledge of a bit of data management and the basics of Lightning is needed too. Easy, right? Go get 'em, tiger!

You have reached a point where you have the certificate and the knowledge. So, how do you start, and which career path do you choose? I think that's an important question, and you have to answer it yourself. But don't worry, I will not leave you with it alone. I will try to help you make that choice. The most important question to ask yourself is whether you are someone who enjoys coding. Is code not a problem for you? Do you like talking to people? And does designing certain technical solutions pose a problem for you?

Why am I asking all this? Because with the answers in hand, you will be able to choose between a few available career paths. Look at the next figure showing career possibilities. Watch out for Pacman.

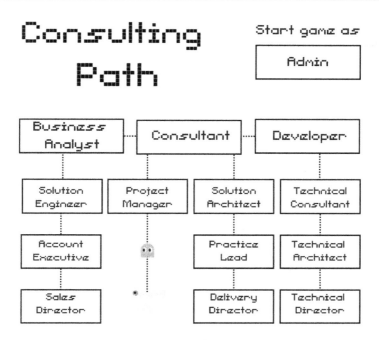

Figure 13.13: Career path

As you can see, you'll begin your adventure by being an admin. Do not try to skip this incredibly important position. You can work as a system admin or as a Salesforce support analyst – the title is not so important. Why not skip this position? As an admin, you learn a lot about the system and discover functions you previously had no idea about. During such work, you will encounter cases written by users. The path of a system admin usually looks like this: you start as a *junior*, then after some time, you move on to the *middle*, and the last title is *senior*, after which it is worth talking to your manager about what is next. This decision will be guided by the direction you want to go and what skills you possess. With these titles, it works a bit like Pokémon – Pichu, Pikachu, and Raichu. So, the final evolution of an Admin is – Senior! I choose you!

Why else start your career as a system administrator? By gaining experience, you acquire information on best practices, interesting solutions, and most importantly, self-confidence. Moving to the next position, you know that you already know something about Salesforce and will not hesitate to use it. Clients can sense when you feel uncertain, but remember that not knowing something is not bad. If you do not know something during a conversation with a client, it is better to say you will check and come back to them with a proper solution, rather than trying to make up something that is not true.

Now, I want you to familiarize yourself with the job descriptions available to you after your admin career. This trio, namely business analyst, consultant, and developer, will help you understand which direction your career might go in. This team is a bit like the characters in Diablo – each has special properties and abilities. Here are our heroes:

- The **business analyst** (**BA**) works a bit like a detective. Of course, we are not talking about finding an art thief or a hacker on the UN's servers. We are talking about someone who perfectly understands the client's business and identifies areas where the business can be improved. The BA works at the intersection of business and IT. They are a bridge between the two worlds of user needs and the technical possibilities of Salesforce. Thanks to their analysis, they can easily draw strategic conclusions from the data, which will help the company modernize its solutions and introduce them to the ever-changing market.

- The **consultant** is like a digital Indiana Jones. They traverse different industries, from which they gather experience and best practices. Their task is to advise on how the system can be used most effectively to achieve the desired goals for the client. It is the consultant who proposes certain solutions because, knowing the system, they can suggest specific solutions. Therefore, when a new project starts, the consultant leads, among other things, discovery workshops, during which they ask many questions and from the answers build the system and solutions, structure, and functionalities. With deep specialist knowledge, they can adapt to a specific environment. The consultant is a kind of guide who takes the company by the hand and walks it through the landscape of objects and records, asking about business needs to finally reach a well-tailored system. What do they do in Salesforce? They translate the client's words into Salesforce language and then work on the system's design.

- The **developer** is a magician of code. In cooperation with the consultant and BA, they can create applications and solutions. The developer can build simple websites as well as complicated IT systems. They mainly follow logic and use programming languages. They are also a bit like an architect and builder in one, as they shape reality in software. What do they do in Salesforce? They build components, processes, triggers, and many other brilliant solutions. This person is technical.

Choosing your future path is a bit like choosing your studies: you may not know what you want to do until the end and then pick something at random (I have a background in physiotherapy and cultural studies). But remember, your choices are not forever. If one day you decide it would be cooler to be a developer instead of a consultant, trust me – the road is open. Working in the Salesforce world, you can easily switch industries. Many people working in the industry started from the simplest tasks, which were repetitive, but through them, they learned a lot about the system and its use. They learned how to navigate it freely and use that freedom in their work.

I'm not speaking only from my perspective – I have also asked other people working in the Salesforce ecosystem for their feedback on how they started their careers and for their golden advice:

* The first one is Piotr Dziedzic, solution architect and Head of Salesforce Consulting. He is a specialist with vast experience and 20 certificates and is generally a Salesforce enthusiast. Here is what he wanted to convey to you:

From my perspective, the most crucial element in developing one's career is finding genuinely cool aspects. Someone might say it is cliché, but this is true: "*Just answer the important question – what do I enjoy doing in life? Then start doing it.*" If learning about a certain area does not bring joy, the entire process will be very challenging. I started with CRM systems around 2015. When I first saw Salesforce and its archaic Classic, I was not convinced (at that time, I was still in love with Siebel from Oracle), not fully understanding how such a system could be helpful at work. As an administrator responsible for data migration, I spent hours clicking through tickets and Excel spreadsheets filled with rows of product descriptions that ultimately had to land in a brand-new Salesforce org. Such work can be demotivating in the long run. However, the consultants working alongside us were leading the entire implementation, coming from a large external company. My curiosity led me to talk with them about what they saw in this system and what their work truly entailed. They explained how crucial data and clarity of processes within the company are. I must admit, their business acumen impressed me – that is when a light bulb went off in my head, and I thought that I would like to have a similar career someday. Over time, comments from users switching to Salesforce also emerged, with people appreciating how work became easier with our cleaned data and enjoying there being less clicking. That is when I realized our work was not meaningless – it might have been laborious and repetitive, but in the end, it genuinely helped these people! This was a major shift in my perception of these tasks, giving me new sources of motivation! That's something I always try to convey to younger colleagues just starting with various systems and administration – tinker with settings.

At the end of the day, it is a real help for someone. My fascination with the work of people who saw a bigger picture of the whole implementation and could find meaningful aspects of value for their client, along with gratitude from end users, made me want to dive in! That was when the Trailhead platform was just starting, with significantly fewer materials than today, yet it was still a real treasure trove of knowledge. I was (and still am) impressed by how many wonderful courses Salesforce offers for free to its community. Working with my sandbox and playing with various setting changes became a habit. It might sound funny, but I often created various scenarios for myself – mainly problems that clients could have – and then slightly role-played being a consultant; I was not only trying to change my system's settings but also thinking about how I would present it to someone on the infamous *call*. I quickly realized that such attempts were fun for me. They brought me more satisfaction than data work or creating user accounts for more people. When I felt more confident in this, I started looking for a new job where I could make use of my newly acquired skills.

After a while, I succeeded. However, while the technical side of my skills developed relatively quickly – completed Trailheads and scenarios greatly aided in gaining the required knowledge – unexpected difficulties were brought by what is called the *soft side*. I am talking about conducting client conversations and business requirement workshops. This side was not "soft" for me at all! I still remember how stressful these public performances were for me. Sweaty hands, shaky voice – it was no different from the stage fright known to everyone. Here, I can say that the way I prepared for such meetings made the biggest difference for me. Instead of going with the flow, I started preparing goals I wanted to achieve at the meetings. Before the workshops, I wrote down a list of questions I wanted to ask. After the meetings, I prepared minutes – I noticed that people sometimes understand some issues differently than I do, and sending a summary allows avoiding such misunderstandings. The biggest difference for me was made by something said by a colleague with more experience in the profession: *"Listen, Piotrek – you don't have to know everything at these meetings. It's really OK to tell these people that you will get back to them with an answer a bit later."* It turned out to be a much better solution than making up answers and sometimes saying things that slightly missed the truth. The sentence *"Let's park this question – I will get back to you on it in a few days"* became a permanent part of my canon. After some time (here again, a cliché – practice makes perfect), I started always having goals, questions, and an agenda for such meetings. Later, I even started trying to learn more about the people I would be talking to before each meeting to adjust the language to their expectations. In my opinion, one of the key skills of a consultant is the ability to convey complex concepts – often technical – in layman's terms and to those who only see our systems from the user interface level. The plus side of working with Salesforce is that there will never be a shortage of knowledge and new things. If you ignite the desire to explore it, there will always be more to learn. Today, after years of working with CRM and in consulting, I am still learning. I've passed 20 certifications, and there are still more to go. We have the privilege, of that working in the legendary "dynamic" environment is in our DNA. Probably the most important advice I can give you is to never stop having fun! For me, this job was never boring again. And I wish the same for you!"

- The next person is Łukasz Bujło, a Salesforce MVP, senior business architect, and the founder of Coffee&Force. He organized Polish Dreamin'24, which turned out to be a brilliant Salesforce event, attended by many guests from Poland and beyond:

"Build your network. Active participation in the Salesforce community is crucial. Join Salesforce Trailblazer Community Groups, several of which exist in Poland, and participate in meetings and Salesforce events. This enables you to stay informed about the conditions in various companies, the current demand for specialists, and to share experiences or receive advice."

"After mastering the basics of Salesforce, focusing on the administrative part, it is worthwhile to think about specialization. Before you do, take a step back and consider your strengths and what you have potential for. For instance, if you are the life of the party and enjoy building relationships with others, you're likely more suited to the consulting path than the developer path. Remember, each path has its unique requirements and offers various career opportunities.

"I recommend considering the Salesforce DevOps path. It's somewhat unpopular and niche, but demand is on the rise. I have encountered numerous examples of individuals who were skeptical about pursuing a developer role but thrived in the DevOps pathway. The need for these skills is constant – even outside of project phases, there is always work involving deployment to production during maintenance stages.

"Theory and Trailheads are important, but practical experience is a different aspect altogether. I suggest taking part in CharITy Hackathons, where over three days, you can see a concise overview of software development from the ground up – from understanding requirements to preparing a proof of concept/demo. I strongly encourage you, either individually or in a team of 2-3 people, to find a small non-profit organization for which you could do an implementation. Salesforce offers 10 free lifetime licenses: with them, the organization benefits from your implementation, you gain experience in requirements gathering and client interaction, and you can implement your first commercial project. It is a win-win situation.

"If you are transitioning into a new career, your previous experience can be a great advantage. If you have worked in the automotive, telecom, or other sectors, your knowledge of business processes and the market is very important. This can make you an excellent consultant and partner in client development projects. Therefore, after mastering the basics, it is worth exploring niche Clouds in the Industries area. There is still a shortage of specialists in these areas, and this could be what sets you apart. Simply knowing Service Cloud or Sales Cloud will not be enough to differentiate you from the crowd and keep you competitive."

This is how individuals who today possess vast knowledge in Salesforce areas started and use it to deliver the highest quality systems to their clients. As you can see, everyone started from lower positions. But hard work and determination bore fruit. As Stan Lee, the creator of many Marvel characters, said, "*Life is not a life if it passes without challenges.*" So, now it is your turn: take on the challenge.

One challenge would be job interviews and conversations with recruiters. As you probably know, LinkedIn is considered the best business portal (not just in my opinion). If you are new to it, here are some of my tips:

- Add recruiters and headhunters from your dream companies, as they might offer you your first job.

- Find Salesforce groups on Facebook and start engaging there.

- On LinkedIn, complete your profile. Do not add information about the job where you handed out flyers on the street; focus on what can add the most value to your profile. Add a description of your work – if you have no idea what to put, look for people who inspire you and, based on them, build your description. If you have worked in IT, describe your responsibilities. It is important for people who want to get to know you.

- Add every one of your certifications and valuable experience. Participated in some volunteering? Add it – people like goodness.

- Add a photo that will intrigue recruiters/headhunters. There are many photos with arms folded across the chest, but unfortunately, fewer interesting photos.

Now, a few words from those involved in recruitment. I asked two recruiters with a lot of experience to share some golden tips on what they pay attention to and what is important for them:

- The first is Bartłomiej Stojak, a recruiter with extensive experience in IT and beyond. He has recruited for many recruitment agencies, not just within the Salesforce ecosystem. Here is what Bartłomiej wants to share with you:

"Salesforce as a platform has an advantage over other ecosystems because it is based on a "live" contact between all its users (this includes those working on its implementations or customizations as well).

"From a recruitment perspective, in the case of Salesforce, I always pay attention to interpersonal skills first and technical skills later. This does not mean that technical skills are not important – it means that certain deficiencies in technical experience can be somewhat compensated for with so-called soft skills.

"If I had to give just one piece of advice, it would be that if during an interview you encounter topics you are not very confident in, instead of responding "*I don't know*" and steering the conversation toward an awkward silence, it's worthwhile to delve a bit into the subject despite a lack of expertise in that area. You could say, for example, "*This is functionality related to X; however, I am aware of the complexity of the topic. It seems to me that it is something similar to Y, but I am planning to dive deeper into the topic.*" Such an approach shows familiarity with the ecosystem and related functionalities, thus leaving the door open for further conversation and making it easy to transition to topics such as your motivation for changing jobs or general developmental plans. By doing this, you demonstrate a willingness to take responsibility for any gaps you may have, but you also emphasize a desire to grow and invest in your knowledge."

Here are a few tips from Phillip Poynton, a seasoned recruiter with a wealth of experience in connecting talent within the Salesforce ecosystem to their dream roles. Phillip has dedicated his career to guiding Salesforce professionals toward opportunities that not only match their skills and ambitions but also push the boundaries of what is possible within the Salesforce community. With a keen eye for talent and a deep understanding of the Salesforce landscape, Phillip has become a trusted advisor to many in the industry.

Preparing your CV and presenting your Salesforce-specific skills and experience

You need to prepare your CV and effectively showcase your skills and experience – here are a few pointers on how to do it:

- **Highlight specializations**: Clearly indicate your expertise in specific Salesforce Clouds (e.g., Sales Cloud, Service Cloud, and Marketing Cloud) and any niche areas, such as CPQ and Einstein Analytics. This specificity helps potential employers quickly identify your suitability for roles requiring those skills.

- **Tailor your CV**: Analyze job descriptions carefully to identify key skills and requirements. Incorporate these keywords into your CV, aligning your experience with the employer's needs. This customization makes your application more relevant and engaging – it can also boost your chances of being shortlisted with applicant tracking systems that automatically screen CVs using specific keywords.

- **Action words**: Use strong action verbs to start each bullet point, such as "implemented," "managed," and "developed," to make your roles and achievements stand out.

- **Quantify achievements**: Detailing your contributions with measurable outcomes (e.g., "Led a team of five in a Salesforce implementation project, resulting in a 30% increase in sales productivity for a customer") offers concrete evidence of your capabilities.

- **Project highlights**: For each relevant role, include a shortlist of projects where you've made a significant impact. Describe your role, the project goal, and the result, focusing on your direct contributions.

- **Showcase learning and certifications**: If you are currently preparing for a certification, mention it to demonstrate your commitment to professional development. List all relevant and acquired certifications to establish credibility. Share your Trailhead profile URL to validate your expertise and certifications.

- **Include a personal statement**: Offer a succinct profile statement that highlights your experience, aspirations, and what you bring to your next role. This personal touch can make your CV more memorable.

- **Digital portfolio**: Provide links to your Trailhead profile, LinkedIn, and any other platform that showcases your projects, GitHub repositories, or personal website. This digital portfolio is crucial for illustrating your practical skills and dedication.

- **Be different – stand out**: As an experienced Salesforce recruiter, I have seen thousands of CVs, but I have only come across a handful that really stood out. One CV I really enjoyed reviewing was designed to mimic a Salesforce Kanban list view and a Salesforce sales pipeline, cleverly mapping the candidate's career journey through the stages of a sales process – a technique familiar to those in the Salesforce ecosystem. This approach not only made the CV visually engaging and thematic but also effectively highlighted the candidate's qualifications and achievements in a way that was both unique and unforgettable.

Utilize your network

Here is how to start effectively leveraging your contacts:

- **Leverage LinkedIn connections**: Use LinkedIn to find and reach out to people working in companies you are interested in. Personal recommendations can significantly boost your chances of being noticed. If you do not receive a response via Linkedin, do not worry – follow up on Trailhead. You know the name of the person you are looking to connect with, right? If so, find them on Trailhead and send them a direct message, which will be automatically sent to their current email address connected to their Trailhead profile.

- **Be open and communicative**: If you are actively looking for a job, make it known in your network. Sharing your journey and aspirations can attract opportunities and support. Maria Śliska and I are currently running a LinkedIn Live series named Salesforce Career Coaching. This initiative offers real-time coaching to navigate the Salesforce job market effectively. We are following Maria Śliska›s journey, a dedicated Salesforce enthusiast, as she aims to land her first job in this field. Our live, unscripted sessions provide insights into overcoming challenges and leveraging opportunities within the Salesforce ecosystem.

- **Innovate your approach**: Stand out by creating a short demo of a Salesforce solution you have developed. This visual representation of your skills can be a powerful tool to distinguish you from other candidates. It is also a fantastic way to showcase your soft skills, which are crucial requirements for all positions.

- **Research and connect**: Investigate companies within the Salesforce ecosystem through AppExchange, LinkedIn, and company websites. Engage with employees and hiring managers to gain insights and express your interest.

Engage with the Salesforce community

Here are a few reasons why engaging with the Salesforce community is worthwhile:

- **Volunteer and present**: Active participation in your local Salesforce community events as a volunteer or presenter can significantly enhance your visibility and network within the ecosystem.

- **Collaborate and learn**: Community involvement is not just about giving; it's also an excellent opportunity to learn from peers and stay abreast of the latest Salesforce trends and best practices.

- **Mentorship**: Seek out mentorship opportunities within the community. Being a mentor or mentee can significantly expand your professional network and learning opportunities.

Maximize networking opportunities

Here is how to make the most of your connections:

- **Strategically attend events**: Research Salesforce-related events and sponsors in your area. Prepare specific questions and aim to learn about open positions directly from hiring managers or representatives.

- **Follow-up**: After networking events, connect with individuals you have met on LinkedIn or via email. A polite follow-up expressing your interest and recalling your conversation can make a lasting impression.

- **Explore various platforms**: Salesforce trailblazers are active across multiple platforms. Join Facebook groups, Discord channels, and other social networks to widen your networking scope.

- **Consult with specialist recruiters**: Reaching out to recruitment agencies specializing in Salesforce roles can provide you with valuable insights and potentially open doors to unadvertised positions.

- **Elevator pitch**: Prepare a concise and compelling elevator pitch for networking events and interviews. It should quickly summarize who you are, what you do, and what you are looking for. Master it! Trust me, the next time you are asked, "*So, tell me something about yourself,*" you will blow them away with your carefully rehearsed elevator pitch!

And that is all from Philip!

Summary

I sincerely hope these recruitment experts' knowledge will assist you in future recruitment opportunities, and that the advice from the first two experts will help you choose the right career path. In this chapter, you've been introduced to the more social side of Salesforce. You've read about people creating their own communities, how to implement your ideas globally, and where to look for events. The last section was incredibly helpful; I wasn't fortunate enough to find a mentor who could give me career development ideas. Thanks to the last section, you can now transform your image in the business network and benefit from the contacts you've acquired. I am confident that after reading this book, you will have substantial knowledge of Salesforce. But do not just stop with this book. Find Trailheads that will introduce you to Salesforce's new features, get to know the upcoming AI, and build your knowledge and portfolio. To conclude this book, all I have to say is this: *May the (Sales)force be with you.*

Index

packtpub.com

Subscribe to our online digital library for full access to over 7,000 books and videos, as well as industry leading tools to help you plan your personal development and advance your career. For more information, please visit our website.

Why subscribe?

- Spend less time learning and more time coding with practical eBooks and Videos from over 4,000 industry professionals
- Improve your learning with Skill Plans built especially for you
- Get a free eBook or video every month
- Fully searchable for easy access to vital information
- Copy and paste, print, and bookmark content

Did you know that Packt offers eBook versions of every book published, with PDF and ePub files available? You can upgrade to the eBook version at packtpub.com and as a print book customer, you are entitled to a discount on the eBook copy. Get in touch with us at customercare@packtpub.com for more details.

At www.packtpub.com, you can also read a collection of free technical articles, sign up for a range of free newsletters, and receive exclusive discounts and offers on Packt books and eBooks.

Other Books You May Enjoy

If you enjoyed this book, you may be interested in these other books by Packt:

ChatGPT for Accelerating Salesforce Development

Andy Forbes, Philip Safir, Joseph Kubon, Francisco Fálder

ISBN: 978-1-83508-407-6

- Masterfully craft detailed and engaging user stories tailored for Salesforce projects
- Leverage ChatGPT to design cutting-edge features within the Salesforce ecosystem, transforming ideas into functional and intuitive solutions
- Explore the integration of ChatGPT for configuring Salesforce environments
- Write Salesforce flows with ChatGPT, enhancing workflow automation and efficiency
- Develop custom LWCs with ChatGPT's assistance
- Discover effective testing techniques using ChatGPT for optimized performance and reliability

Salesforce for Beginners

Sharif Shaalan, Timothy Royer

ISBN: 978-1-80323-910-1

- Explore business development with leads, accounts and contacts in Salesforce
- Find out how stages and sales processes help you manage your opportunity pipeline
- Achieve marketing goals using Salesforce campaigns
- Perform business analysis using reports and dashboards
- Practice automating business processes with Salesforce Flow
- Gain a high-level overview of the items in the administration section
- Grasp the different aspects needed to build an effective Salesforce security model

Salesforce DevOps for Architects

Rob Cowell, Lars Malmqvist

ISBN: 978-1-83763-605-1

- Grasp the fundamentals of integrating a DevOps process into Salesforce project delivery
- Master the skill of communicating the benefits of Salesforce DevOps to stakeholders
- Recognize the significance of fostering a DevOps culture and its impact on Salesforce projects
- Understand the role of metrics in DevOps architecture within Salesforce environments
- Gain insights into the components comprising a Salesforce DevOps toolchain
- Discover strategies for maintaining a healthy Salesforce org with a variety of supporting DevOps tools

Salesforce Platform Enterprise Architecture

Andrew Fawcett

ISBN: 978-1-80461-977-3

- Create and deploy packaged apps for your own business or for AppExchange
- Understand Enterprise Application Architecture patterns
- Customize the mobile and desktop user experience with Lightning Web Components
- Manage large data volumes with asynchronous processing and big data strategies
- Learn how to go beyond the Apex language, and utilize Java and Node.js to scale your skills and code with Heroku and Salesforce Functions
- Test and optimize Salesforce Lightning Uis
- Use Connected Apps, External Services, and Objects along with AWS integration tools to access off platform code and data with your application

Packt is searching for authors like you

If you're interested in becoming an author for Packt, please visit `authors.packtpub.com` and apply today. We have worked with thousands of developers and tech professionals, just like you, to help them share their insight with the global tech community. You can make a general application, apply for a specific hot topic that we are recruiting an author for, or submit your own idea.

Share Your Thoughts

Now you've finished *Salesforce CRM Administration Handbook*, we'd love to hear your thoughts! Scan the QR code below to go straight to the Amazon review page for this book and share your feedback or leave a review on the site that you purchased it from.

`https://packt.link/r/1835085695`

Your review is important to us and the tech community and will help us make sure we're delivering excellent quality content.

Download a free PDF copy of this book

Thanks for purchasing this book!

Do you like to read on the go but are unable to carry your print books everywhere?

Is your eBook purchase not compatible with the device of your choice?

Don't worry, now with every Packt book you get a DRM-free PDF version of that book at no cost.

Read anywhere, any place, on any device. Search, copy, and paste code from your favorite technical books directly into your application.

The perks don't stop there, you can get exclusive access to discounts, newsletters, and great free content in your inbox daily

Follow these simple steps to get the benefits:

1. Scan the QR code or visit the link below

https://packt.link/free-ebook/978-1-83508-569-1

2. Submit your proof of purchase
3. That's it! We'll send your free PDF and other benefits to your email directly

www.ingramcontent.com/pod-product-compliance
Lightning Source LLC
Chambersburg PA
CBHW080623060326
40690CB00021B/4793

* 9 7 8 1 8 3 5 0 8 5 6 9 1 *